A Construction Guide for Yorkie in 32r

Designed by Derek Crookes

For Parts Supplied by Model Engineers Laser

By Model Engineers Laser

Northstar Support Publishing

1 3 5 7 9 10 8 6 4 2

Northstar Support Publishing, www.northstar.support

Copyright © Model Engineers Laser, 2021

The author has asserted their right to be identified as the author of their Work in accordance with the Copyright, Designs and Patents Act 1998

This edition first published in the United Kingdom in March 2015

Details in this guide were accurate at the time of publishing

A CIP catalogue for this book is available from the British Library

ISBN 978-1-8381132-6-1

Contents

Dedication and Thanks 6
About Model Engineers Laser Limited (MEL) 7
About Laser Cutting 10
The Tools You Will Need 12
    Hand Tools: 14
Equipment Required for Running 16
    Running 17
About Yorkie 21
    The Model 22
Building 23
    The Frames 23
    The Wheels 25
    The Bearings 26
    The Axles 27
    The Cranks 27
    The Coupling and Connecting Rods 28
    Making the Pins 29
    Making the Motion Bracket 31
    Making the Expansion Links 32
    The Motion Work 33
    The Cylinders 34
    Cylinder End Covers 36
    Steam Chest 37
    The Cross Head 38
    Assembly of the Motion Work 40
    Pump and Eccentric 41
    Boiler 43
    Boiler Fittings 48
    Belpaire 49
    Smokebox 50
    Cleading 50

- Bypass Valve and Pipe Work 51
- The Tanks 51
- Reach Rod and Links 54
- Gas Tank, Lubricator and Burner 54
- Pipe Work 56
- Plate Work 57
- Radio Control 60

Painting 60
- Suppliers 63

Drawings 64

Appendix One 108

Dedication and Thanks

Our thanks go to Peter Hawkins who worked at the company and supplied the drawings,

Malcolm High who wrote the first edition.

Second edition revisions by Edward Parrott.

About Model Engineers Laser Limited (MEL)

Model Engineers Laser started back in 2005 when Malcolm High first retired. Over one Christmas he was hacking out frames for his Holmside using a hacksaw, chain drilling, and finally, the milling machine. A few months later, Malcolm was introduced to a laser cutter for the first time, and realised there was a much easier way to cut metals, and not only that, it would be considerably more accurate.

Malcom started producing files for some of the popular locomotives for friends, and then friends of the friends, and then fellow model engineers who didn't even know the friends of the friends Model Engineers Laser was born! It was intended to be a one day a week venture, as something to keep Malcolm occupied. Fast forward fifteen years to 2020 and Model Engineers Laser could boast a product list of over 22,500 parts on the database, had sent out over 425,000 parts all over the world, some of which to places as far away as Tasmania. With cuts of about 1500 parts being arranged nearly every fortnight, ranging from tiny O Gauge parts, to 10 ¼" gauge, as well as large custom cuts, it was no longer the one day retirement job that Malcolm intended, and it was time for Malcolm to retire, properly this time.

Malcolm was very careful to ensure that the new owners were also model engineers who had the right skills to take the business to another level. Some of the customers have some very technical questions, and then there are the 1500 parts that are cut every fortnight that need identifying, sorting and distributing. The new owners would need to have a good understanding of engines and what different shapes are, as they really are just shapes, with no indication of what they are. The new owners really need to have some idea of what they are looking at, but not only that, Malcolm had built this business himself, and he wanted to ensure that he was handing it over to worthy recipients, who would look after it, his loyal customer base that he had spent so much time looking after, and, of course, have the best interest of the hobby at heart.

In 2021, Edward Parrott purchased Model Engineers Laser and it ticked all of Malcolm's boxes: it was the perfect fit having been a customer for over 10 years, a turner by trade, using CAD and CNC machines, but skilled in other engineering disciplines too. He is an active member of his local Model Engineering society and also 3rd generation volunteer on the Talyllyn Railway as a Blockman and Fireman. Both his, and his local Model Engineering Societies' models featured extensive use of laser cut parts, all of which had been produced by Model Engineers Laser. Edward saw a great opportunity to take on the amazing range of products, and could also see further development opportunities too.

MEL continue to offer the bespoke service cutting, and all the original catalogue of parts, but continue to grow it on a very frequent basis. Edward's wife Holly helps with the business too, having come from a business background, but also a modeler in her own right, and a volunteer on the Talyllyn Railway since 2005. They make a great team and quickly obtained a reputation for a friendly team who provide great customer care, in depth knowledge and a cost effective service.

Laser cutting offers, for many model engineers, a fast and accurate way to progress and finish a project. Its popularity within the hobby has increased dramatically. MEL understands the difficulty modellers have in getting laser cut parts that don't cost a fortune, so trying to keep prices down without compromising service is really important. Part of this is by keeping overheads low, to pass that saving onto customers. MEL do this by working from home, reusing packaging where we can (also, this is really great for the environment), we are also not doing this to make a huge profit, but as a service provider for a hobby the owners love. We want the hobby to survive, have affordable options for all budgets to enjoy, and to meet like minded people in our customers at shows and events.

www.modelengineerslaser.co.uk
sales@modelengineerslaser.co.uk
07927 087 172

About Laser Cutting

Laser cutting works by melting, burning or vaporising the material, while an assist gas is employed to "clear" the cut zone of the molten / burnt material or the gas vapour. In the early days of laser cutting, the setting of the laser to produce the desired effect was very much a manual process and very complex. The latest machines now come with many of the common parameters pre programmed, allowing much easier setting. However, operators still require many hours of training to run a laser safely, efficiently and economically.

Laser cutting is a technology that has been used in industry since the 1970's. The first common application was for sign making, mainly cutting acrylic. Since then, the process has been adopted and adapted by virtually every industry group, and is now a significant process in every manufacturing economy. Laser cutting is excellent at processing many different materials. Unlike physical machining laser profiling is not affected by the hardness of a material, meaning materials such as high carbon steels can be cut as easily as standard mild steel.

The cutting process is very complex, but basically involves pre piercing the material outside the area of desired cut, moving the laser beam into the cutting area to apply heat, and finally use an assist gas to remove the heated material and produce the cut. The type of assist gas employed is critical, and is dependent on the material to be laser cut; most commonly used are Oxygen (used predominantly for carbon steels), Nitrogen (used for nonferrous steels & non metals) and Argon (used for more exotic materials such as titanium).

Introduction

This building guide is not intended to be used as a step by step process on how to build the model, but as its title suggests – a guide to aid to fellow model engineers, especially in areas where innovative techniques have been used. We also do not supply all the parts/list them, for example: most engineers have ⅜" brass bar, hanging around in the corner with all the other bits we keep that will come in some day. If not, they are readily available from many places.

Laser cut parts offers many advantages, primarily it produces parts that are very accurate in their profile. This allows for a slot and tab technique to be utilised for construction methods. Slot and tab construction is a proven, fast and accurate process that is especially important in the construction of valve gears, all of which are laser cut.

Another advantage is speed! Model Engineers Laser started when the founder Malcolm was frustrated with the laborious task of cutting the frames out for his own engine by hand. The process of cutting the part for model engines by hand would take years, with laser cutting this can reduced significantly. Not only this, with modern technology there is an increasingly high chance of achieving success first time.

To make this model, a small workshop is required with a list of the minimum tooling given later. For the beginner (or those who have made a mistake or can't wait to get their model done) some of the more complex parts can be bought ready built, for instance the cylinders, pump and boiler. Fortunately, these very reasonable price and are increasing in popularity.

The Tools You Will Need

Anyone with young people in their lives will know the frustration they have when they get a much loved and wanted toy, to find out that you don't have the right batteries. Building this engine is much the same, even though many of the components for this engine can be purchased already completed, there is still some work to do. To avoid your own 'wrong batteries' moment, ensure that you have the correct tools before you start.

Some of the parts are silver soldered together, and if you have never done it before it may seem rather daunting at first as you need to get the parts you are working on up to a temperature of over 650 degrees Centigrade. I can assure you that this is not difficult when the parts are relatively small and a suitable blow lamp can be purchased from most DIY shops. For some of the finer soldering a 25 to 35 watt iron is ideal, with a resistance soldering iron being a great all rounder for this scale modeler as it only heats up a small area and by using soldering cream, the joints can be made very neatly.

Silver solder: get the lowest melting point with a 0.7mm diameter wire. Using a 100 watt soldering iron should be fine and not too powerful for the solder, and to stop silver solder adhering to parts you don't want, simple cover the section with Tipp ex.

Fluxite or Multicore solder, as used for small electrical items and printed circuit boards.

Drilling is required and a drill post is preferable, however, if this is not available then a small pistol type drill will do if care is taken to get the holes square (this is important and will prevent you from cursing a lot later down the construction line. Some holes are tapped ready and 8BA is the most commonly used, feel free to use the metric equivalent, but they are readily available to get should you need to buy them.

If you are going to turn the bushes and other parts yourself, then a small lathe is required. There are a number on the market costing a couple of hundred pounds up to a thousand or so. Suppliers do provide bushes and if you do not want to do this operation you will find someone who can supply them at a reasonable price, but remember, you tend to get what you pay for, so the cheapest option may not be the best choice and you can also explore the second hand market, but be careful to take someone with your who knows what they are doing before you buy. Make sure the lathe comes equipped (or can easily be obtained), collets can be worth their weight in gold and a vertical slide is handy if you do not have access to a milling machine.

A building frame is not essential, but can make life a lot easier, the one seen in the picture here has a 32mm Yorkie frame in. These build frames are readily available from MEL for engines up to 5" gauge and can be built in a day. The

Aluminium extrusion is also used in the rolling roads (available up to 7 ¼" Gauge), you can keep your build frame for your next project, or you can get the extra parts from MEL to convert the Aluminium extrusion into a rolling road for your finished model.

If you do have access to a milling machine, it can make your project go a lot easier, or, as mentioned, you can use a vertical slide on your lathe. If you do use a mill, it doesn't have to be a large one as the parts are only small.

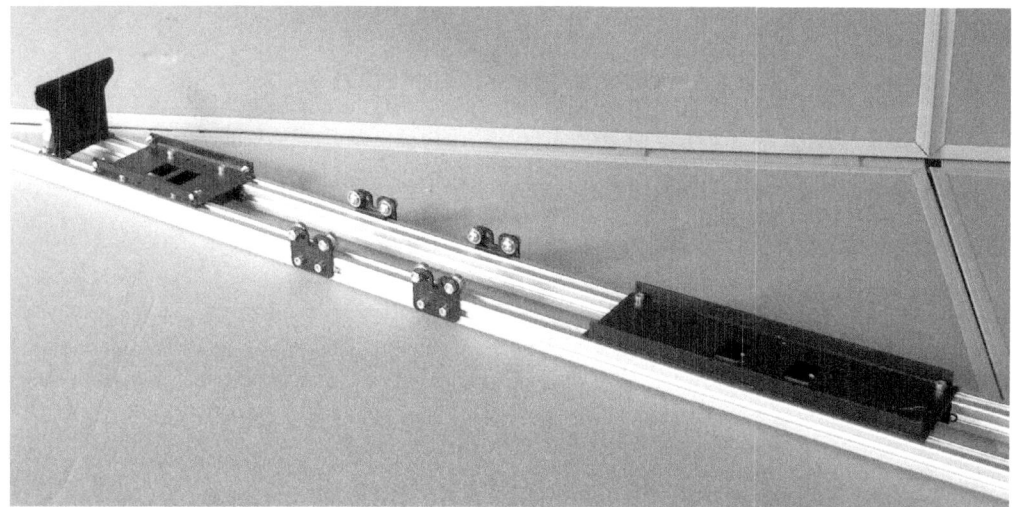

Hand Tools:

You will need an assortment of hand tools and measuring equipment. We are dealing in fine tolerances with these locomotives so a good micrometre is essential. Digital micrometres are now priced very reasonably and certainly make the job much easier. Other than that, a steel rule and good quality marking out equipment should be to hand.

Set of files: purchasing quality files is getting extremely difficult and you will need a range: small needle files from 1.5mm thick up in various shapes and sizes, and for larger work some 6" files for cleaning up the slots and edges on the parts.

Other than that, the usual hacksaws – main frame and junior – plus the fine razor variety will be required.

There are a number of holes to tap and rods to thread. These are generally BA sizes but you can substitute the metric equivalent if you so wish. The most common taps and dies will be 8 and 10BA. These are relatively small so be careful, it is easy to break a tap but not so easy to get it out again!

There will be more than the above list but it is a start.

So far as materials are concerned you will need some brass angle. This is available from model shops and should be easy to obtain.

Equipment Required for Running

Yorkie is a gas fired locomotive, and use either butane or a butane propane mix. When you are making your decision on what to run your engine, it is worth noting that at some events they only allow butane engines.
To light the burner, a gas lighter is the best option

To fill the boiler with water, the best option is a hand operated garden spray with a plastic tube 4mm OD adapted to it. To do this remove the front nozzle on the spray and turn up an adaptor in brass to fit your tube. You may have to open out the hole in the front of the spray. Put the adaptor through the hole and fasten it back onto the spray. Attach the tube to the adaptor.

Steam oil is used in the lubricator and any light oil is suitable for oiling the motion etc.

Blowers: to help you get up steam when lighting your engine, they run on batteries, but some places have points for blowers to plug in to. MEL also sell blowers ranging from garden gauges to 7 ¼" gauge.

Running box: Many modelers complain about finding suitable running boxes for their equipment. MEL have created 2 running boxes, the smaller drawer type aimed at smaller gauge modellers as the height in the back compartments will fit tall gas/butane bottles/methane etc. Larger drawer type is aimed for 5" gauge+ modellers. They are designed to fit other things such as oil cans, batteries, little tins, and blowers too.

Running

Boiler water really needs to be soft water. Having tried distilled water, suitable for batteries, in a 5" gauge loco, it still seemed to get salt deposits in the boiler so now use filtered rain water.

What you need is a rain water butt; a diverter is placed in the down pipe which are available from most hardware outlets. Near the bottom of the butt is a tap to take off the water. The water will be very dirty so a filter is required, these can be obtained from DIY stores and they generally filter down to five microns. It was easier having made a stand and a bowl for filling; the outlet empties straight into a 5 litre plastic container. Using a bowl is slow, to speed up the process, couple it directly to the water butt tap as shown and now it takes around 5 minutes to filter a gallon of water.

The boiler is filled through the 4mm Enots connector, a cheap pump can be made from a garden sprayer. Remove the front nozzle, don't lose the spring and make up the brass adaptor shown in the drawing. Open out the hole in the nozzle to 4.2 mm and push the brass adaptor through it, now push the 4mm plastic air pipe onto the brass and you are done. Commercial versions of this unit are available.

It takes approximately 100 ml of water to fill the boiler. If you have fitted the electronic water level indicator, you can couple this up to a receiver battery and as you fill the boiler the red light will go out and then the green light will come on indicating that it is full.

The lubricator needs filling with steam oil and is designed for 90w grade, thinner grades are available from various suppliers but cannot be guaranteed to work with the lubricator. Open the drain valve on the bottom of the lubricator and with an oil can, fill the lubricator from the top until full. Close the valve and fit the screw top. Continuing on the oil theme, all the bearings require some light machine oil.

There are two types of gas normally used on 16mm locomotives. One is pure butane, the other is a butane propane mix. The latter has more calorific value but is not allowed by some places, check before you fill the gas tank. To fill the tank you need an adaptor, these are available from a number of 16mm suppliers. To fill the tank, fit the adaptor to the gas cylinder and turn it upside down. Ensure the gas tap on the tank is closed and then press the adaptor down onto the Ronson filler on the tank. It takes some force to get the liquid gas into the tank so make sure you hold Yorkie firmly. When liquid gas comes out of the filler it is full.

The burner is lit via the chimney with a gas lighter which can be obtained from most of the large supermarkets. They can be filled with the same gas and using the same adaptor as the gas tank. Turn the gas on at the tank and allow it a second or so to clear the air out of the gas pipe and jet. Now light the gas lighter and hold it over the chimney, the burner should light, however, sometimes the flame does not go back to the burner, it burns at the end of the fire tube. If this is the case reduce the gas flow by turning the valve towards off until it "pops" back to the burner. The note will change to a roar when it does this. The jet has an extremely small hole and is easily blocked, especially when using new tanks that have lots of debris from the silver soldering operation in them. If this happens the jet has to be removed from the holder and blown back using compressed air.

If the burner lights successfully, it will take around 10 minutes to get the boiler to its working pressure - make sure the regulator is closed or it will never get there!

Put the valve gear in full forward by moving the reach rod into its front slot, open the regulator and probably nothing will happen! The cylinders will be full of condensation so turn the wheels in a forward direction to release the excess water through the chimney. The cylinders will then clear and the engine will run. Take it slowly at first, then open the regulator a bit more to increase the speed. Finally, close the regulator until the wheels are just turning over. It may well be a little stiff at first but after some running it will run fine. Do not run it for too long, ten minutes initially is long enough, especially if you do not have a level sensor fitted - you do NOT want the boiler to run dry.

About Yorkie

The prototype was built by the Yorkshire Engine Company at their Meadowhall works in Sheffield. The site can still be seen from the M1 motorway. The order was placed in 1915 by the Indian State Railways' United Provinces Public Works department. Due to the hostilities, the locomotives were not delivered until 1919.

The order was for three locomotives, which were to be used on construction work. They were to be capable of hauling 80 ton loads up a 1:100 incline at 8 mph., all on 18lb rail.

In running condition, they weighed in at twelve and a half tons. The boiler delivered steam at 150psi to the 8 inch cylinders and the centre drivers were flangeless so they could negotiate the tight radii.

These were the only two foot gauge locomotives the company built; being works numbers 1283 7 and it is not known when they were scrapped.
For a more detailed history of the Yorkshire Engine company see the book by Tony Vernon (ISBN 978 0 7524 4530 4).

The Model

This is intended to be a quick build for either the first time builder or the more experienced builder who wants something to put on the track quickly. It is hoped it will not be too expensive and every effort has been made to ensure the cost is kept down. Most of the bar material will be in the "Will be useful one day" drawer. If not, there are plenty of suppliers, some of which are in a list at the end of the book.

The use of laser cut parts is extensive. There are also a number of novel methods of construction that make the build accurate and fast.

The main frames are available as a TIG welded construction. If you are going to build your own frames, I have included a section on this.

Building

Where we can, MEL designs are created with a slot and tab method of construction. It is strongly recommended that you have a trial run before finally assembly.

The Frames

The stretchers must be put in the correct position and that the horizontal one needs some filing as the vertical one that goes with it is at a slight angle - check this against the drawings. Assembly is straight forward if you take care and the frames can be silver soldered together nicely.

Below are listed a number of suggestions:

TIG welding is the preferred method of construction: If you are unfamiliar with this method, please refer to the article in the December 2011 edition of Engineering in Miniature. It is a very versatile method suitable for many types of material and a wide range of thicknesses. However, unless you have used this before it will probably not be your preferred method.

Silver solder is something most model engineers are familiar with - use the lowest melting point available and the correct flux. The positive side to silver solder is its strength; the frames should certainly not come apart again. The down side is the risk of distortion and the cleaning off of the flux. Regarding distortion: to reduce the risk you can either heat the frames with a large flame to 400 degrees, then work from one end or using a smaller flame do it one joint at a time. Whichever method you choose the parts can be held together whilst soldering by damaging the protruding part of the tab with the gentle application of a centre punch.

Soft solder is probably not a method many would consider but for these frames it is suitable. As already stated, strength is not really an issue and with a melting point in excess of 140 degrees there are not many instances where the solder would melt. Steel can be difficult to solder but the use of solder cream makes it very easy. Assemble the frames with solder cream in the joints and along the edges of the stretchers. A small flame or heavy-duty soldering iron of over 100 watts should be adequate. If additional solder is required multicore electrical solder is eminently suitable.

There are many different types of adhesive on the market. These range from the cyano quick sets to the epoxy two packs with a possible longer setting time. New adhesives are coming on the market all the time. There is a LokTite that has small pieces of rubber in the thin cyano mix which allows some flexibility in the joint, making it less liable to fatigue fractures. These adhesives are all suitable for temperatures up to 100 degrees so temperature is not an issue.

The Wheels

The wheels have to be turned to the 16mm association standard. The current recommended profile is shown in the drawings. Take a piece of 1½ inch diameter free cutting steel and chuck it in the three jaw. Face off and turn down the first three quarters of an inch to a diameter of 1.374 inches. Face off another 1/8" to leave the central boss half an inch in diameter. The tread and flange can now be turned; to achieve the three degree taper you will need to use the top slide. Centre drill the hole and drill a hole three sixteenths in diameter through the wheel blank. The hole can now be finished off with a 7/32nd machine reamer. Finally turn the 0.9 inch recess, this is not essential but it makes for a far better looking wheel. The wheel is now ready to part off, only another five to do! The middle wheels sets are flangeless, just as on the prototype.

On smaller lathes it may not be possible to part off this diameter of bar cleanly. If not, an alternative method is to rough out the wheel so that it is about ten thou over size and do not put the taper on the wheels or the flanges. Cut the wheel off by using a saw or other means. Put the wheel in the three jaw and face the back off to the finished thickness. Do all six wheels like this and then make a mandrel. Bolt each wheel in turn to the mandrel and turn the taper on the flanges and treads.

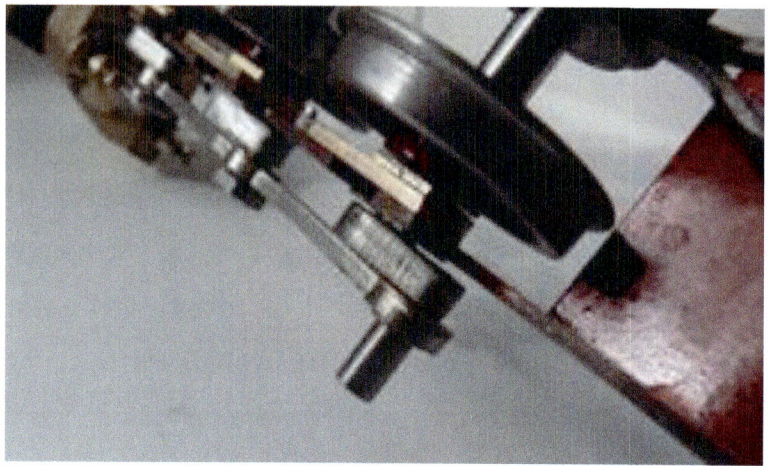

The wheels can now be painted .

The Bearings

The bearings are made from half inch diameter brass and are simple turning jobs so should not create any problems. Place the brass in the three jaw, drill and ream the hole. The slot can be cut with a parting tool, use the hole in the frames as a guide for diameter and width. The flat is best put on now, either in a mill or with a file. It is easier to hold the bearing whilst it is still attached to the main bar, rather than later when it is much smaller. Finally, return to the lathe and part off - you will need six in total.

The Axles

The axles are where collets come in, take a piece of ¼" diameter steel bar cut to the length shown in the drawings and turn one end down to 7/32". If you have collets, you can now turn the bar round and do the other end with confidence that both will be concentric. One of the bearings can be used as a template to ensure the final diameter is correct. If you do not have collets, then it may be easier to use a piece of 7/32" diameter bar and fit a piece of ¼" overall diameter tube over it as a spacer, allow for this when you make the pump eccentric.

The wheels can be attached to the axles with Loctite, using a high strength retainer. Make sure that wheels are in the correct location on the axles, slide the bearings onto the axles and fit into the frames. The bearings are held in the frames by axle box keeps, these are tapped 10BA, the corresponding holes in the frames are drilled 0.072 and then countersunk.

The Cranks

The fly crank is made from 5/32" or 4mm steel, once profiled, the holes have to be positioned accurately. If you use the parts supplied by MEL rather than making your own, then our laser cut parts are spotted. If you are making them yourself it will need some careful marking out, ensuring that the two holes need to be perpendicular to the faces, if done with care, this can be done on a drill post. The axle hole has to be reamed 7/32", the other hole is tapped 6BA. It is possible to use one of the cranks as a jig to drill the other five.

The crank is attached to the axle with a 6BA grub screw, an Allen socket is preferred. The hole can be drilled in the drill post if care is taken.

The eccentric crank is from 1/16" plate, the stub axle is turned from 3/32" diameter bar as shown in the drawings. This needs to be done in one setting if you are using a three jaw chuck, to attach it to the crank, we would use silver solder.

The Coupling and Connecting Rods

The rods are made from stainless steel; this ensures they will not rust. The laser cut rods need the edges draw filing to a finish and the holes cleaning out, as there may be a small nib to remove.

Bushes are made from either phosphor bronze or brass and can be fixed into the rods by Loctite or an appropriate soft solder. The bushes are a simple turning job but do require the holes reaming to a finish.

Make up the six pins and fit the wheels, cranks and rods to the frames. The pins for the rear wheels which also take the connecting rod are different from the rest as they have to take the crank. The cranks will be quartered later.

Making the Pins

There are a number of pins to make for the rods: take a piece of round bar and put it in the lathe, use collets if you can (they are not expensive and you are guaranteed concentricity). Turn down sufficient length to the diameter required for the bush plus the tapped portion, then turn down the tapped portion to the correct diameter.

The important part: now to tap only the portion required, do this by placing the bush on the pin and then tapping the rod until the die touches the bush. The lead in on the die will then give the clearance required.

If you do not have a tail stock die holder you can use the method shown in the image. On the tailstock chuck, open the jaws so that they are fully inside the chuck. Place the die in its die holder up to the rod and move the chuck up so it touches the back of the die holder and start to turn the holder. Keep the chuck up against the die as it progresses, thus keeping the thread square to the job. The coupling rods can now be fitted: do one side at a time, fit three cranks loosely on the ends of the axles, then fit the coupling rods. With the cranks at their lowest position, tighten the grub screws. Turn the axles through ninety degrees and fit the cranks and coupling rods on the other side, again, at their lowest position. A jig for doing this is shown above and is available as a laser cut part. Your wheels should now turn freely.

Making the Motion Bracket

MEL's motion bracket is a slot and tab kit, it is very important that it ends up square and there is a right and left hand. Start by identifying the parts on the sprue and cut through the steel holding the parts together, so you end up with five pieces. Leave as much material on the tabs as possible and clean out the slots with a needle file, so the tabs will go through with a little force - they do not want to be slack. When you are confident the parts are right, clean them up and assemble the bracket. Damage the protruding tab so that the parts will not fall apart during soldering. Check it is square, flux up and using a low melting point silver solder and heat the component in a hearth. It is relatively small so it will not need a big flame.

Short pieces of silver solder can be put onto the bracket where required or you can just go in with a stick when the bracket is hot enough, it is up to you. To clean the bracket dip it in weak acid, citric from a home brew shop will do but is often not out on the shelf and you will need to ask for it. Clean it up, file off any excess, and open out the holes if required to the dimensions shown on the drawing. Soft solder the bush for the expansion link as shown.

Making the Expansion Links

These are another laser cut MEL kit which comes in two parts plus a pin. The parts are going to be bolted together and then silver soldered. Start by lightly centre popping the area between the bolts on the 1.5mm thick piece, this will ensure there is a small gap for the silver solder to penetrate. Make the pin and thread 8BA for 1.5mm and tap the holes in the expansion link back plates to accept this thread.

Clean the parts up and bolt together with steel bolts, preferably old rusty ones. Tipp Ex can be applied to the areas you do not want soldering. Apply flux and heat the components up to silver solder them, make sure the pin stays perpendicular to the plates. Once done, immerse in acid to clean up.

Remove the excess bearing plate material so it looks like the drawing, this can be done with a junior hacksaw and a file. Part the two links and file off any excess, and to finish up, tap the pin and clean out the hole for the eccentric rod pivot.

The Motion Work

All the forked ends of the motion work are made in the same way, bolt together and silver solder. To get the four thou gap required for the solder to penetrate, the area concerned has to be filed away. First, identify where the silver solder has to be applied. In these areas, four thousandths of an inch should be removed with a file. Apply flux and bolt the parts together with rusty steel bolts, you can apply Tip Ex to the bolts if you want to be certain the solder will not adhere.

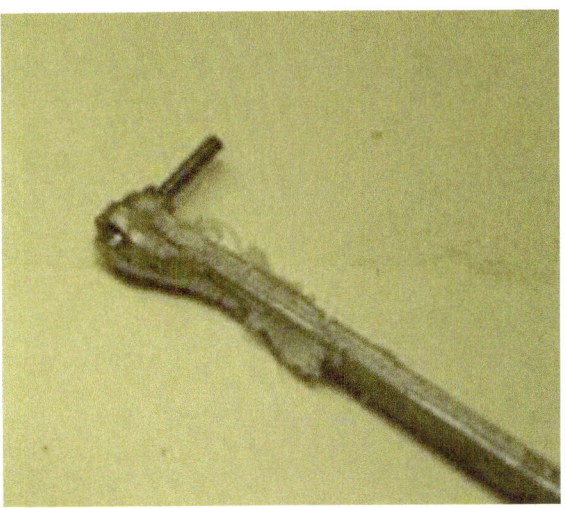

Apply more flux, heat up and apply solder, always using the lowest melting point silver solder you can get, clean in acid, remove the bolt and clean up. The centre of the fork can be removed in a mill with a slitting saw or with a junior hacksaw. Clean up with a file and open the hole out to the required diameter. On the radius rod the pin that engages the expansion link needs to be a neat fit.

The Cylinders

There are options with regards to material

One inch square phosphor bronze is available from a number of suppliers: this material can be difficult to tap and there are a number of 10BA holes to tap.

One inch square brass is available from many suppliers at reasonable cost: it taps easily and should give many years of reliable running.

Start with a piece of one inch square material 1" long, face off 1/8" in the lathe to get the 7/8" dimension. If you are not using the jig, face off both ends in the 4 jaw, do not face to length yet. Blue up one end and on the marking out plate, mark the centre of the bore and use a centre punch to mark it accurately, as it is to be used with a centre finder to position it in the 4 jaw. If you are using the jig, place the brass in the jig and mark the centre of the bore with a drill. When assembling the jig just peen the ends of the tabs over do not solder as this may leave a cusp in the corners.

Return the block to the 4 jaw and using a centre finder accurately centre the cylinder ready for boring. Start with a centre drill, follow up with a 3/8" drill then open out with a boring bar to just 20 thou in diameter under the 1/2" final bore diameter. Run a 1/2" reamer through the bore and finish off with a hone. Finally, before removing the block, face off to length.

Place the block in the jig, make sure you know which holes require drilling, what diameter they are and how deep they have to be. Check twice, drill once as there are differences that hand the cylinders. If you are not using the jig you need to mark out the holes accurately. Tap the holes as shown on the drawing.

The final holes are the steam passages from the cylinder to the ports. To do this, the cylinder needs to be held at an angle in the mill, this is for a small flat to be cut and the holes can then be drilled. Use a small centre drill first to mark the spot and then follow up with a sharp 1/16" drill. This is nearly the final operation on the cylinder block and is probably the most difficult so take your time!

Finally, the bottom of the cylinders has to be rounded. This is best done on the milling machine. If you do not have one, then mark out carefully and remove as much of the material as you can with a hacksaw before finishing off with a file.

Cylinder End Covers

The front cover is the easiest to start with and can be turned from a piece of standard bar. The rear cover is much more complicated as it has the piston rod through it and it holds one end of the slide bars. Start with a piece of stock bar and turn this down to the outside diameter, then turn the bar to make the seal housing. Drill for the piston rod, then open up for the stuffing box as shown on the drawings and tap the hole. Use an end mill to produce the flat bottom to the hole. Use a parting tool to reduce the diameter of the bar where it is to enter the cylinder. It is important that this diameter is accurate so take care! Finally, part off the end plate.

It is now necessary to finish off the mounting for the slide bar and tidy up the end plate. If you are using the jig this is made easier by first drilling and bolting the endplate onto the jig using the four bolt holes. Put the jig in the mill with the plate bolted to it. Mill the flat for the slide bars to bolt to, it is important this is the correct distance from the centreline of the piston rod. Dimensions are shown on the drawings.

The piston is turned from a piece of stock bar, final turning of the diameter can be done on the piston rod. Turn down to within 10 thou in diameter, drill and tap for the piston rod before parting off.

Steam Chest

This is made from a piece of 1" x 1/4" brass bar 1" long, on the mill, cut out the centre as shown and then use the jig to bore the holes. Take care drilling the holes as this hand the cylinders.

Make the gland for the valve rod, the steam chest cover is a laser cut part and only needs boring as shown on the drawings. It is extremely important to get the valve dimensions correct, a laser cut kit is available from MEL for this or you can make it out of the solid. The valve rod is a length of stainless steel rod threaded as shown.

Assembly of the cylinders requires a number of gaskets, make these from 20 thou gasket paper, however, they are available as a laser cut part. Use graphite yarn or O rings for the piston ring and the gland stuffing.

The Cross Head

This comes as four separate laser cut parts, two of which are the same. Identify the parts and clean them up, cut the drop link and drilling template off the sprues and put them in a safe place. Remove four thou from the 5mm thick inner in the areas shown in red on the drawing. Coat the areas in blue with Tip Ex.

Bolt the parts together and fit the top, the tabs can be gently peened to stop it falling off. The two drop links are fitted to the same side with small bolts or rivets. These need some Tip Ex on to stop them being soldered, flux up and silver solder.

The areas marked in blue have to be removed on the inner side, leaving the drop link proud of the cross head, this can be done by filing or on a mill.

Cut the two cross heads off the sprue, clean up all the edges with a file and clean up the slot for the slide bar and check it moves freely. Fold the drilling jig and fit over the cross head. To align, push the slide bar through the rectangular holes. The 8BA piston rod hole must be perpendicular to the face, this needs some careful alignment to get it right. Use the slide bar as a guide, the slide bars bolt to the cylinder and motion bracket.

Assembly of the Motion Work

The cylinders, rods and motion work can now be assembled. Refer to the general arrangement drawing to see where all the parts fit. Make sure the expansion link can move sufficiently, the corners need to be filed as shown in the drawing to achieve this.

To set the crank correctly it needs to describe an eleven millimetre diameter circle. To set it up, put the piston in the middle of its stroke and measure from the cylinder to the centre of the crank and record this. Turn the wheels through 180 degrees and measure again, the difference should be eleven millimetres. If it is not, adjust and do it again until the measurement is correct.

Set the piston in the middle of its stroke and the radius rod in mid gear. At this point the slide valve should be covering the ports and a slight movement of the radius rod should uncover them. That is all there is to it, couple up to an air supply and it should run.

Pump and Eccentric

The feed pump is a simple turning and silver soldering operation, to seat the balls, tap them with a light hammer into place.

The eccentric strap and rod come as laser cut parts. Carefully part the eccentric straps and file the edges to clean off the cutting marks. The mating faces need to be flat so start with a fine file and finish off on a flat surface with a fine emery cloth.

It is best to soft solder the two parts of the strap together; this ensures the parts do not move during drilling. Once together, clean off any excess solder and mark the holes for the 10BA bolts. This needs to be done with some care as there is very little clearance and drill the holes 1.4mm making sure the parts are vertical in the drill vice. Clean the hole out at regular intervals or you risk breaking the drill.

Once the holes are drilled mark the edges of the strap so that they will go back in the same position, heat up the soft solder and part the strap. The surfaces will need cleaning up again, do this carefully so as not to remove too much material. Tap the holes in the rod and open the holes in the strap out to 10BA clearance (.069") and bolt together with 10BA bolts.

The two rods are bolted onto the strap: Drill out the holes and tap one of the rods and bolt the rods onto the straps. Locknuts may be a good idea or some screw lock, the rods need parting slightly to go around the pump rod.

To make the eccentric itself place a piece of 5/8" bar in the 4 jaw offset 0.94". Drill and ream the 1/4" hole. Now, transfer the bar to a 3 jaw or centre it in the 4 jaw. With a parting tool, turn the slot in the bar as shown in the drawings.

Test with the eccentric strap until it is a running fit and drill and tap the 6BA for the grub screw. Finally return to the lathe and part off.

Boiler

The boiler is a very simple affair, those who have built in the larger scales will be used to having flanged plates, tubes, crown stays, firebox stays and combustion chambers. Well, there are none of those in this boiler. Basically, it is a tube in a tube with 3mm end plates. Very simple, but effective.

Whenever copper tube is held in the lathe it must have some form of support inside it, so first turn down two pieces of wood around two inches long to fit in the larger tube. Whilst in the lathe, centre drill one so it can be supported in a centre in the tail stock. It is best to leave a slight collar on this piece so it does not disappear down the tube if any pressure is put on it.

After cutting it to length, blue up the outer tube where it will require marking out, this does not have to be too accurate. A long scribe is required down the length of the boiler for the safety valve and the dome retainer. With a scribe held in the tool post at centre height, the tube in the 3 jaw and the other end supported in a centre, mark a line along the length of the boiler. This will be the datum line. Two more holes are required at ninety degrees to this line 0.4 inches back from the front of the boiler. If you have a digital inclinometer this is easy, as all you have to do is attach it to the chuck and move the latter round until the desired angle is reached before marking out again. If you do not, then a ribbon of paper can be put around the tube and cut so it is the same length as the circumference, divide this length by four and mark the paper again. Put the paper back around the tube with one of the marks lining up with the scribed mark on the boiler. Two of the other marks will now be at ninety degrees, drill the four holes as shown on the drawing.

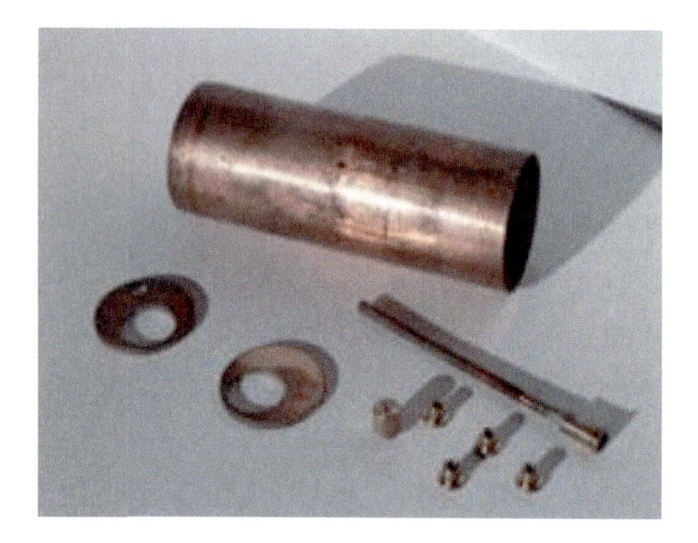

The inner fire tube can now be cut to length, the two end plates are in 3mm copper. Start by carefully marking out the plates, then by chain drilling or maybe on a band saw, cut the discs out. The fire tube holes will have to be finished by boring out in the lathe.

There are a minimum of five gunmetal bushes to make, two are for the water feeds, one is from the pump and the other is the manual feed. Another is to fit the dome; one is for the safety valve and the final one is for the regulator. These are all simple turning jobs, if you are going to have water level probes, then there are two more to make for the back head, both tapped 3/16" x 40. Alternatively, you can just tap the 3mm copper plate if you decide to fit probes at a later date. The boiler is now ready to assemble.

If you have not made a boiler before, here are a few tips:

First, please refer to the series of articles on silver soldering by Keith Hale which were published Engineering in Miniature. Generally, cleanliness is king when it comes to soldering, for this an acid tank is required. This does not have to be very big but needs to be of stout plastic construction. Citric acid is the easiest to obtain, go to your local home brew shop and get some crystals. Make it up as instructed on the packet, you will need some strong plastic gauntlets and a plastic hook or a set of long grips to lift parts back out of the bath.

A length of stainless steel rod around 2mm in diameter makes a good poker, sharpen the end. You will also need a small hearth, this can be a few fire bricks arranged on a steel topped table, it does not have to be complex. For heat, plumbers' gas torches are too small. The standard Sievert gas burner 2941 gives out up to 7.7 Kw of heat; this is more than enough for our purposes. You will need some suitable flux and low melting point silver solder, 1.5mm diameter is ideal.

Start by having a dry fit of the parts, there needs to be a gap of around 0.1mm around the contact area so that the solder will penetrate. It is easiest to solder one end at a time, it does not really matter which you do, out of preference, we start at the smoke box end. Clean everything up in the acid, wash off and place the outer shell on the bench. Push the two bushes for the clack valves through the holes in the shell and now slide the fire tube into place and both end plates. Make sure the fire tube is square with the end plates and place vertically in the hearth with the smoke box end uppermost, the two bushes should stop the end plate from falling through. Place bricks around the end to be soldered to reflect the heat and protect the surrounding area. Mix up some new flux with a little water until it makes a soft cream. It can be useful to have some short lengths of silver solder in the area where you are silver soldering.

Cut 3 4 lengths of solder, approximately half an inch in length placed around the joint between the shell and the end plate. Cover in flux.

You need the appropriate safety equipment; some clothing is not fire proof; buy yourself a decent set of welding overalls and some heat resistant gloves. It is also a good idea to wear safety glasses in case the flux splatters, it is highly corrosive. Small fire extinguishers can be purchased cheaply, but can save you a lot!

Start by heating up the area around the boiler, the flux will start to bubble. Have your poker and a new length of silver solder easily to hand. Once the flux has started to go like glass the solder will start to melt. The poker can be used to gently coerce any lengths of silver solder that try to move away from the joint. Once the solder starts to melt, add more solder from the full length. Solder the two bushes in using the full length of silver solder. When complete, remove the heat and allow to cool before dunking the boiler into the acid bath. The copper will be annealed, so do not drop it in or you are likely to damage the end. Check around for any obvious faults before starting on the cab end, which is done in the same manner. Finally, the dome retainer can be soldered in, this can be done with one of the ends if you feel confident enough.

Clean out the threads and it is time for testing! These boilers are below the Bar Litre minimum specified in the pressurised container regulations; however, you would be very irresponsible to fire up a boiler that has not been tested. The 16mm Association recommend that all boilers are pressure tested. First, make sufficient plugs for each of the openings in the boiler. You will need one adaptor to go between the pump and your boiler, check with your club boiler inspector what thread they use on their test equipment.

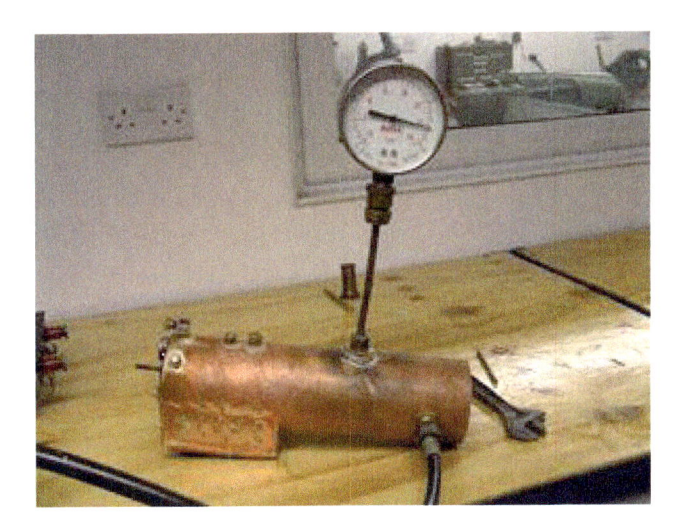

The boiler can be filled from the test pump by leaving one of the plugs out, the air has to go somewhere. Once water is seen at the bush the plug can be refitted and the boiler pressurised up to twice its working pressure, so 80psi (under current regulations at time of publish). It has to be left there for 10 minutes, during this time you can check for leaks. Some form of numbering is required on the boiler for identification purposes. If you have a small set of number stamps this can be done on the outer shell opposite the end plate.

The end plate is also a possibility but be careful as the copper will have annealed.

Boiler Fittings

The safety valve is a simple turning job, there is a pipe going from the valve through the roof. This will be described later and although it looks a daunting task, it is in fact a fairly simple and rewarding operation.

The regulator is a simple turning job, too. There is insufficient room for a glass, so level sensors are recommended, the two clack valves are another turning and silver solder job. From the rear the left hand one is for the axle pump and the right for a hand pump, there is not a lot of clearance between the tanks and the smoke box.

Pipes run back from the clack valves to the cab, electronic water level indicators are available, in place of the sight glass.

Belpaire

In order to keep the boiler simple, the Belpaire was not included, this is a dummy and comes in three parts. Clean out the slots and form the outer as shown in the drawings, push the tabs through the slots and solder the unit up. Make sure the hole for the safety valve is towards the end with the square cut outs. It is probably best to silver solder, a higher melting point soft solder will suffice. File the corners off to complete the unit.

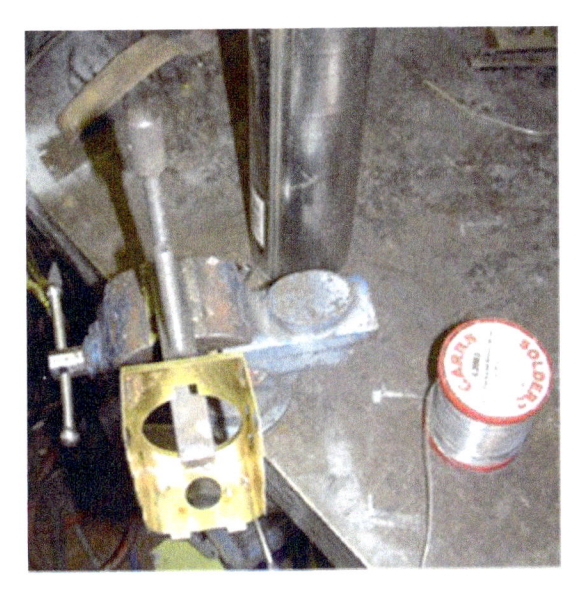

To fit the boiler, push it through the front aperture upside down. Once the safety valve bush is in the box, turn the boiler through 180 degrees. There is not a lot of room to do this and you may have to lightly file off the edges of the safety valve bush but the minimum required as the valve itself has to seat on this.

Finally insert the boiler in the rear hole.

Smokebox

This is formed from three laser cut parts, a front, a rear and a wrapper. Remove the chimney support piece from the smoke box front and form the wrapper around a piece of suitable bar until the tabs will enter the slots. Clean everything up, push the tabs through the slots and with an old 10mm bolt, fit the chimney support. This fits inside the smoke box in line with the hole for the chimney. Ensure it is parallel to the base, silver solder together, remove the bolt and clean up in acid.

The chimney and petticoat pipe can be turned freehand from stock bar.

Cleading

Make the cleading from 10 or 20 thou shim brass, cut to length and form this around a piece of tube, similar in diameter to the boiler. Insert the holes for the dome and clack valves.

The simplest form of insulation is to use masking tape, with boiler bands that can be made from 0.5mm brass, 2mm wide. This is not a normal stock size but Model Engineers Laser can provide suitable material. Put the insulation around the boiler followed by the cleading, clamp up and fit 3 boiler bands, these can be soldered on their ends to hold in place.

Bypass Valve and Pipe Work

The bypass valve is turned from stock and silver soldered together. The feed to the valve comes from the outlet of the pump. When the valve is closed, the water is directed to the boiler, when open it goes back to tank.

The Tanks

The tanks come as laser cut parts, the left tank has the water in and the right tank is for the radio receiver if fitted. Start by riveting the tank sides, it is recommended that MEL's riveting tool is used for this with the jig. There are left and right hand tanks and they are different.

There are some positioning tabs on the jig, these have to be pushed through until they are just proud of the plate. Bolt the tank side to the jig with M4 bolts and the anvil spigot needs to be 1mm high and 1.5mm in diameter. Do not drill a central hole, it is not required - the spigot is only there to locate the jig.
Set the depth gauge on the riveter so that it creates a rivet of suitable height. Form all the rivets in the side panel, you will have to keep moving the bolts around to gain access to all the rivets. When completed push the tabs through to the other side and do the other tank.

Form the curved corners on the tank and the bunker by hand, the position and diameter of the forming bar is shown on the drawings. Use a piece of soft wood to protect the rivets. Clean out all the slots and make sure the parts will go together easily, start by soldering together the base, side and end plate of one of the tanks, then fit the inner plate, the end is bent through ninety degrees on this plate.

Fit the base for the bunker and then the inner plate, again the end of this has to be turned through ninety degrees. The top hand rail can now be fitted, make up the tank support bracket and the tanks are ready to go on the locomotive.

Reach Rod and Links

The reach rod stand is best silver soldered, this also takes the micro servo if you are fitting radio control. The lifting links are silver soldered together and cleaned up. The final filing for clearance may be required on these. The lifting arms and reverse arm are laser cut parts silver soldered onto brass bushes as shown.

The weigh shaft can now be fitted with the links and reach rod. The end of the reach rod is turned through ninety degrees to give better access.

Gas Tank, Lubricator and Burner

The gas burner is a piece of 3/8" thin walled brass tube that is widely available. Carefully mark out and drill the holes as shown in the drawing, turn the blank end from brass bar and silver solder in place. The other end of the burner tube fits into a piece of brass turned as shown in the drawings. This needs to be a tight fit in the fire tube, the other end is turned and tapped to hold the gas jet. None of these turning jobs should present builders with any problems.

The gas cylinder sits on the left side of the footplate, it's made from a piece of 22mm diameter tube. The ends are pieces of brass bar turned as shown in the drawings, the base has a tapped spigot to fasten it to the footplate. The filler is a Ronson type which has a very unusual thread; M4.5 x 0.5. Taps can be obtained from a number of suppliers. On the filler there is a small hole on the threaded section, this is the vent and it must not be covered up. The tube and the end plates are silver soldered, the cylinder must be tested to 300 psi, and needs an identification number stamping on it. On top of the cylinder is a gas valve, this is turned from brass bar and the individual parts are silver soldered, the handle is from 3mm diameter stainless steel.

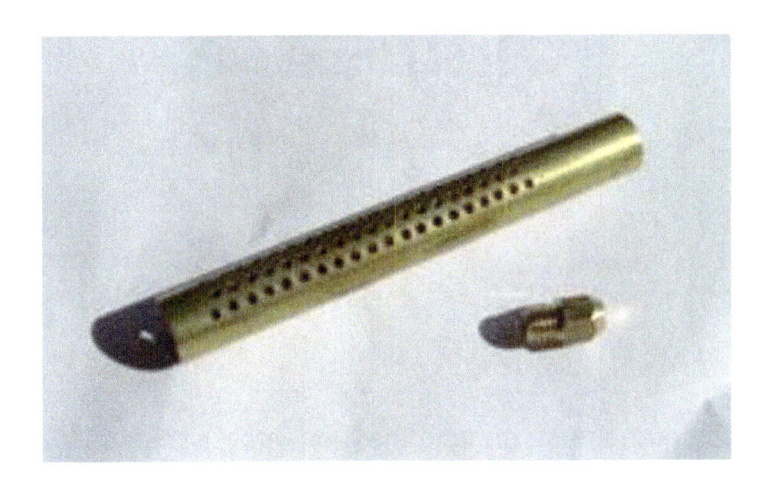

In this scale, lubricators are of the displacement variety, this is another tube with end plates so is probably best made with the gas cylinder. On the bottom plate is a drain; this pipe is also used to fasten the lubricator to the footplate. Steam is fed through the 1/8" diameter copper tube in the top of the lubricator. Any steam that condenses displaces some of the oil which is taken through into the cylinders via the 0.2mm diameter hole in the pipe. The lubricator is filled via a screwed cap in the top plate. Since the lubricator is at steam pressure it requires testing to twice boiler pressure.

With the lubricator and gas tank bolted to the footplate the pipe work can now be coupled up. A pressure gauge is fitted off the regulator and needs to be in a visible position when the locomotive is in steam.

Pipe Work

From the regulator the steam goes to the lubricator, it then drops down to the super heater which is threaded though the burner holder. This pipe is coupled to the inlet manifold as shown in the drawings.

For manual filling, a 4mm commercial ENOTS straight connector is fitted to the clack valve on the right hand side of the boiler. This is fed from a hand spray with a piece of 4mm pipe adapted to the nozzle.

The pump is fed from the left hand tank, which then goes to the bypass valve. When the bypass valve is closed, water is fed to the boiler via the left hand clack, when the valve is open the water goes back to tank.

Plate Work

There are a couple of plate work items to complete. The cab footplate is a piece of 1mm brass. There are a number of holes to drill in this as shown in the drawing. It is bolted to the rear horizontal plate on the buffer beam. There is also the front footplate to make which attaches to the corresponding plate at the front of the loco - neither of these should cause any problems.

The roof is another piece of 1mm brass formed as shown in the drawing. Holes are required for the safety valve exhaust pipes and the gas filler. You can leave the latter out if you wish but it means removing the roof every time you have to fill the gas tank. It is connected to its supports by four sockets as shown in the drawing, these are a simple turning job and can be attached to the roof with soft solder. If you are going to use radio control, the batteries will end up in the roof, we could not find any other position for them

The roof is supported on four 1/16" diameter rods mounted in 3/16" square brass as shown in the drawing. Start by cutting four pieces of brass bar to approximate length then face off to final length in the lathe. If you have a self centring four jaw, that is ideal, otherwise, you will have to centre it manually. One end of the rod is tapped 10 BA to accept the bolts that hold it to the footplate, the other is drilled 1/16". It may be that you can only get 1.5mm diameter stainless rod for the uprights, if this is the case, drill 1.5mm. Mark out; drill and tap the 10 BA cross holes, these attach to the cab plates, so if you have the laser cut parts these can be used as a template. Finally cut the stainless rod to length and Loctite it into the holes in the ends of the brass bar.

The back plate is another piece of 1mm brass, it is not known if there were any rivets in this plate so it has been left blank. If, in the future this information becomes available, then the correct rivet pattern can be produced. The only real difficulty on the back plate is the narrow strip that sits on the top. If you are using the laser cut parts, the strip has slots in whilst the back plate has corresponding tabs. This ensures the joint is not only strong but that the top plate sits centrally along the back plate. The back plate has two of the roof support sections bolted to it and these, in turn, are bolted to the footplate, thus fixing the back plate in position.

The safety valve vents through two pipes which go through the roof. Start by turning the base from a piece of one inch diameter brass.

The pipes are made in two sections, the bottom section is a piece of ¼" copper tube formed as shown. To stop the tube collapsing when being formed, it can be filled with lead or soft solder. Melt this down and pour into the tube sufficient to fill the section being formed. Alternatively, obtain some round plumbers' solder that will go into the tube and make a slight bend in the tube where the final bend will be. Thread the solder into the tube and melt it with a small blow torch, so that it fills the area you have bent. Once formed, heat the tube again and allow the lead or solder to flow out. Alternatively, make a steel former out of a piece of one inch diameter bar. Cut a groove in the bar to accept the copper tube such that the inner diameter of the bar is now a quarter of an inch. Anneal the copper tube and gently form it in the jig. On completion, the bottom of the radius will need removing in order to silver solder it to the base you have already turned. Cut the copper to length and soft solder in place the two pieces of brass tube to complete the safety valve vents. These need to line up with the two holes in the roof.

Radio Control

2.4 GHz is the preferred frequency, the receiver goes in the right tank, the battery can be secured under the roof. A micro servo is used to control the regulator. The switch can be mounted on the frames in the hole provided.

Painting

Yorkie is a very simple engine and I am not going to go into great detail about how to paint your model, our choice is to use aerosol cans where possible.

We find them easy to use and compared to an air brush, much more reliable as there are no worries about getting the mix right, the right consistency and being able to repeat the process when you need to refill. Cleaning is another issue, with an aerosol you just invert it and spray until the paint has gone from the head, with any form of air brush the whole gun needs to be cleaned. It is also much easier to obtain aerosols in any colour you want from your local auto paint supplier, at a reasonable cost.

It is not known what colour Yorkie was out shopped in, as we have no colour photographs. From the image available it was probably grey primer. So, when the locomotive got out to India it could have been painted in any colour that was readily available. The choice then will be yours, but whatever colour you choose, the loco will need to be completely stripped down for painting.

A self etch primer is essential as a base coat, and for the steel parts a standard primer from your local car accessory shop is normally adequate but for the brass components you will need a special adhesion aid. Both are available in either spray cans or tins for air brushing. Prepare all the surfaces well and give the paint a key by using a 400 grade paper over the surfaces. Clean off any particles using tack cloth and then a suitable solvent. Do not try to put on too thick a layer, two coats are usually better than one. The brass adhesion aid we have used is transparent, so the second coat on the brass components needs to be a grey primer to give the top coat some backing colour.

The amount of pigment and thus the covering qualities of top coat paint varies greatly, not only between paints but also between colours. I tend to find that reds and blues cover better than yellows and greens. It is possible to get base colours, these cover better but will be dull on completion, so need a coat of lacquer. At the end of the day the choice is yours, go with what you are confident with.

If you wish to line the tanks, a lining guide is available in laser cut Perspex. They can be attached to the tanks with blue tack and use a suitable lining pen. Another option for lining is to print them onto water transfer paper, this is available on the internet from various places. The lines can then be transferred the model as water slide transfers.

The only hot surface is the smoke box and this needs a special paint. No primer is required, just a good key on the steel. The paint is usually stable up to six hundred degrees centigrade and can be purchased from auto paint suppliers in aerosol format.

Suppliers

(Correct at time of publication)

Brass etc.

Blackgates Engineering

Unit 1, Victory Square, Dewsbury, WF12 7TH (01924) 466000

www.blackgates.co.uk

Brass rod, Steel

Noggin End Metals

83, Peascroft Road, Norton, Stoke on Trent, ST6 8HG, 01782 865428,

www.nogginend.com

Silver Solder and Solder Paint

CUP Alloys

15 Sandstone Avenue, Chesterfield, S42 7NS, 01246 566814,

www.cupalloys.co.uk

Gas fittings

Roundhouse Engineering

Churchill Road, Doncaster, South Yorkshire, DN1 2TF, 01302 328035

www.roundhouse eng.com

Drawings

The following drawings are for Yorkie in 32mm Gauge, designed by Derek Crookes.

YORKIE

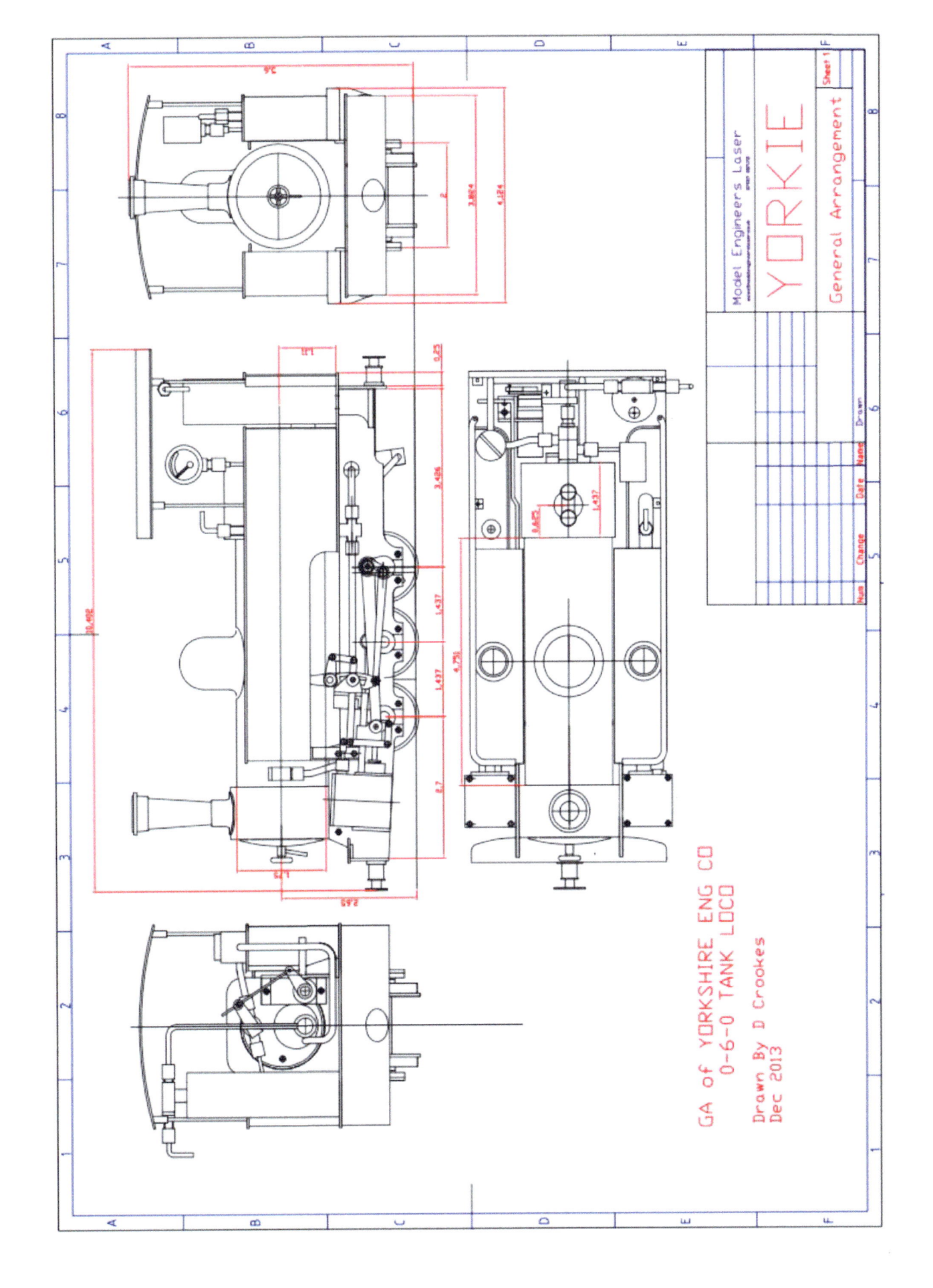

Model Engineers Laser

General Arrangement

Sheet 1

GA of YORKSHIRE ENG CO
0-6-0 TANK LOCO

Drawn By D Crookes
Dec 2013

YORKIE

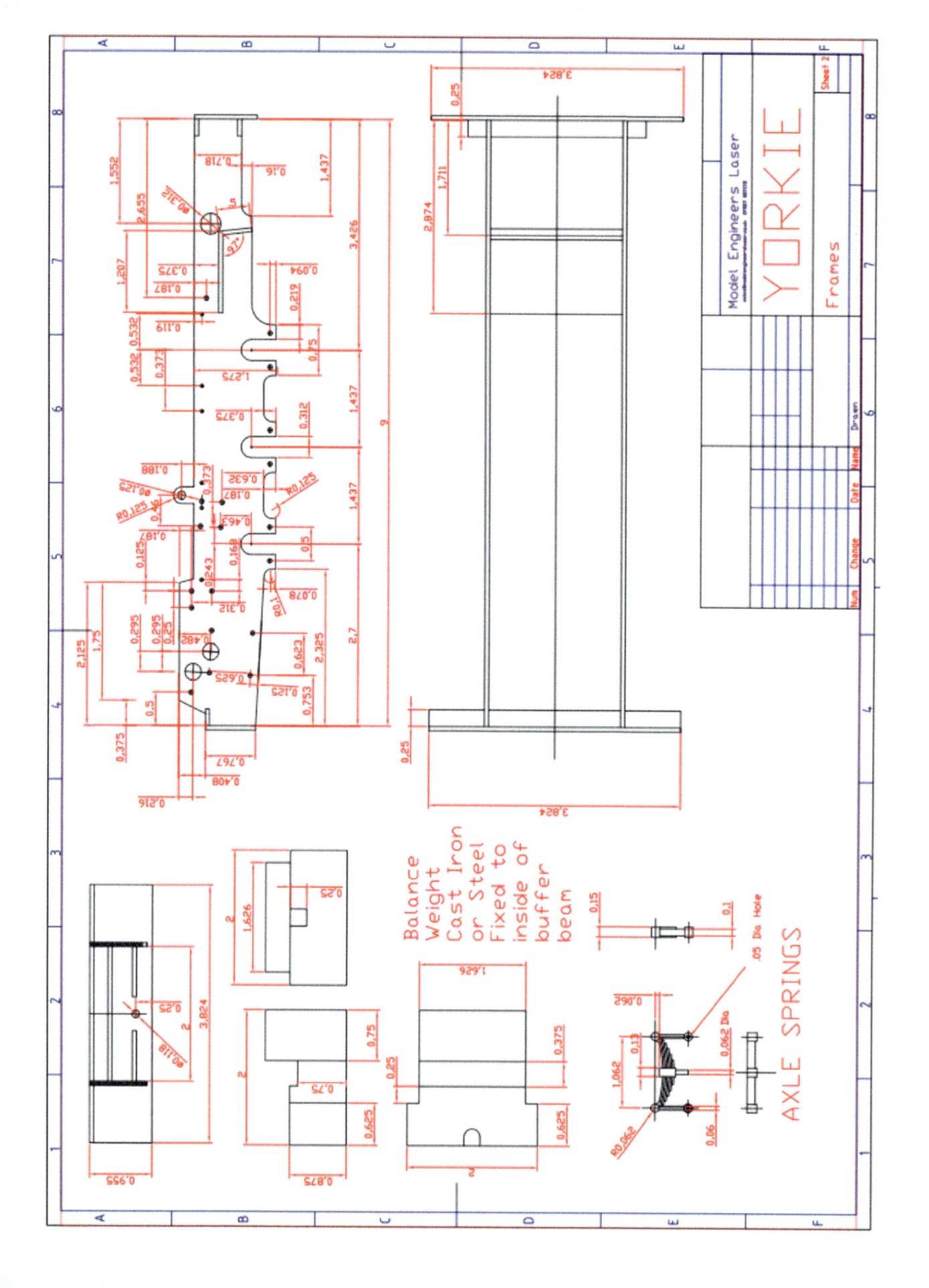

Frames

Balance Weight Cast Iron or Steel Fixed to inside of buffer beam

AXLE SPRINGS

Model Engineers Laser

Sheet 2

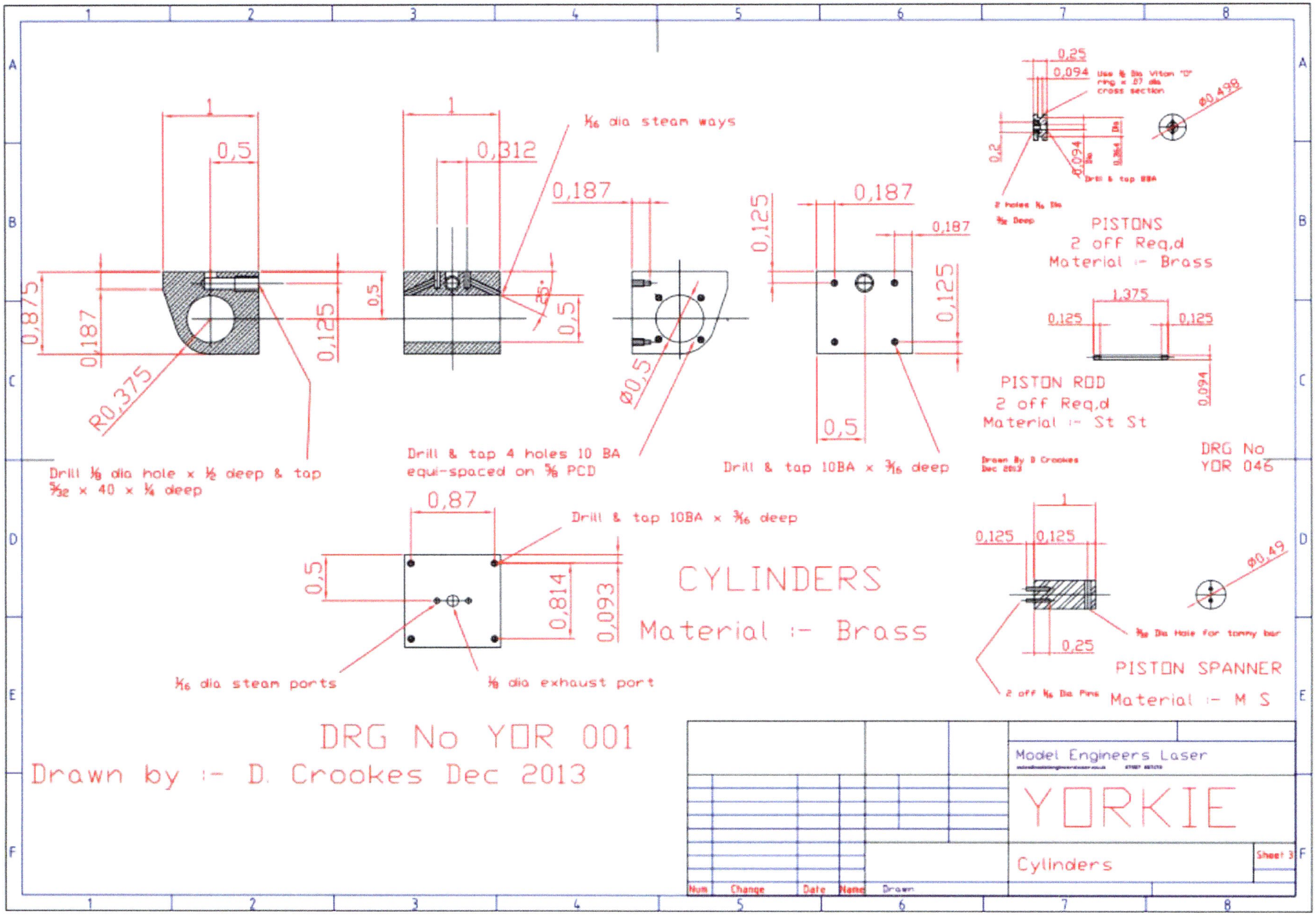

1/16 dia steam ways
Drill 1/8 dia hole x 1/2 deep & tap 5/32 x 40 x 1/4 deep
Drill & tap 4 holes 10 BA equi-spaced on 5/8 PCD
Drill & tap 10BA x 3/16 deep
Drill & tap 10BA x 3/16 deep
1/16 dia steam ports
1/8 dia exhaust port
CYLINDERS
Material :- Brass
PISTONS
2 off Req,d
Material :- Brass
PISTON ROD
2 off Req,d
Material :- St St
DRG No
YOR 046
PISTON SPANNER
Material :- M S
2 off 1/16 Dia Pins
DRG No YOR 001
Drawn by :- D. Crookes Dec 2013
Model Engineers Laser
YORKIE
Cylinders
Sheet 3

STEAM CHEST
Material :- Brass

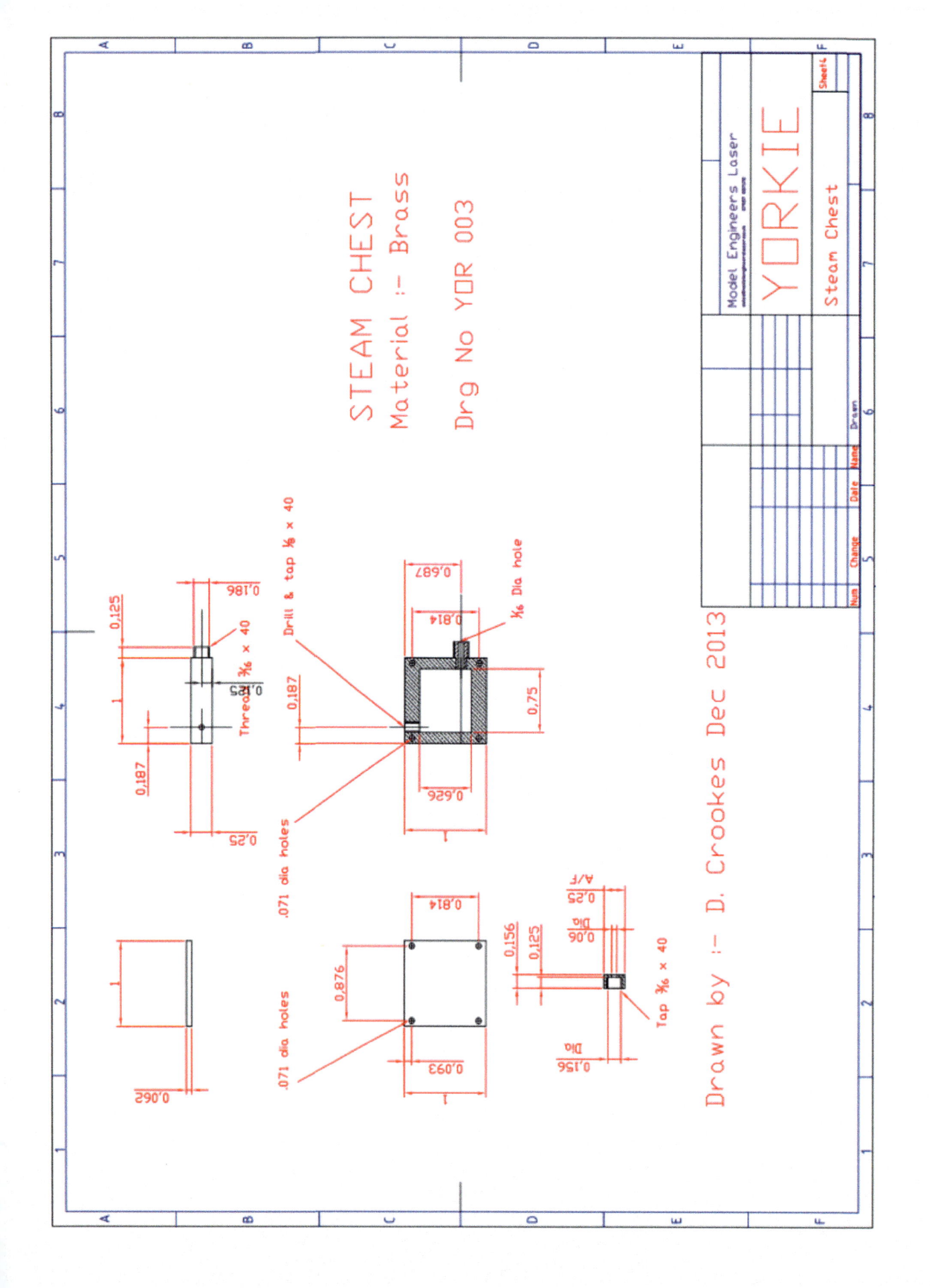

Drg No YOR 003

Drawn by :- D. Crookes Dec 2013

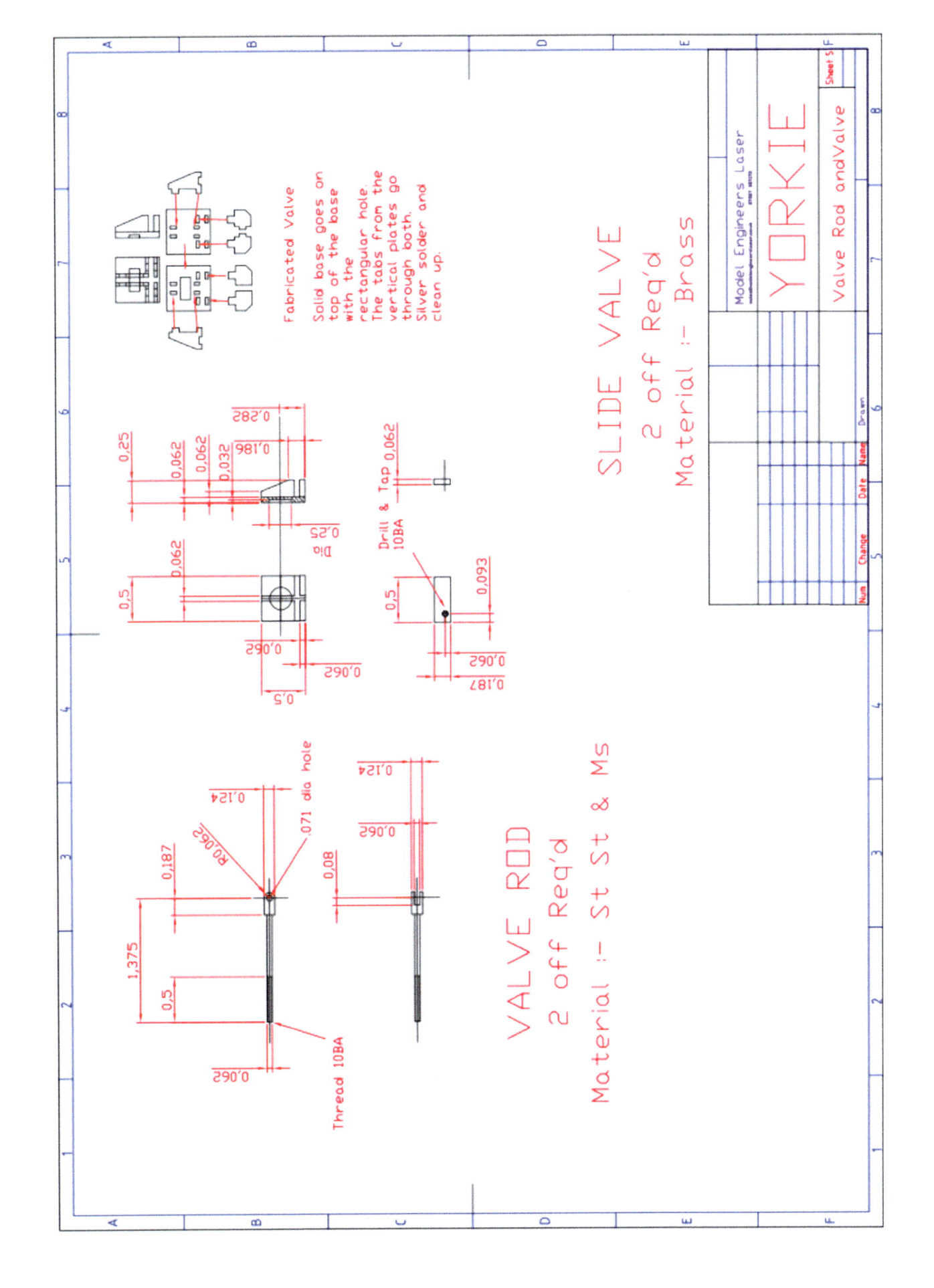

SLIDE VALVE
2 off Req'd
Material :- Brass

Fabricated Valve

Solid base goes on top of the base with the rectangular hole. The tabs from the vertical plates go through both. Silver solder and clean up.

Drill & Tap 0.062 10BA

VALVE ROD
2 off Req'd
Material :- St St & Ms

Thread 10BA

.071 dia hole

R0.062

YORKIE

Valve Rod and Valve

Model Engineers Laser

Sheet 5

CYLINDER END CAPS

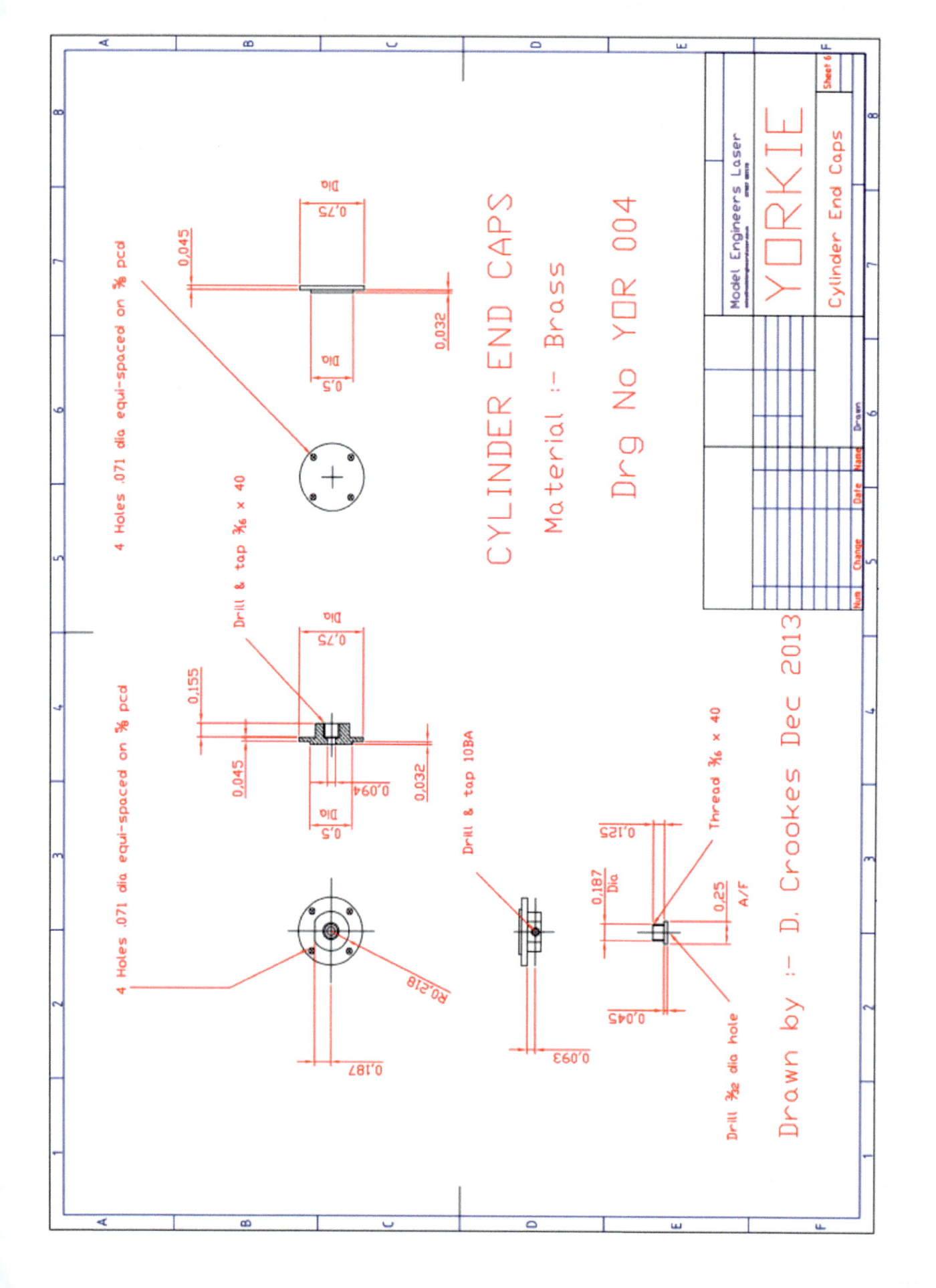

Material :- Brass

Drg No YOR 004

Drawn by :- D. Crookes Dec 2013

- 4 Holes .071 dia equi-spaced on ⅝ pcd
- Drill & tap ³⁄₁₆ × 40
- 0.045
- 0.75 Dia
- 0.5 Dia
- 0.032
- 0.155
- 0.094
- Drill & tap 10BA
- R0.218
- 0.187
- 0.187 Dia
- 0.125
- 0.25 A/F
- 0.045
- 0.093
- Thread ³⁄₁₆ × 40
- Drill ³⁄₃₂ dia hole

YORKIE
Model Engineers Laser
Cylinder End Caps
Sheet 6

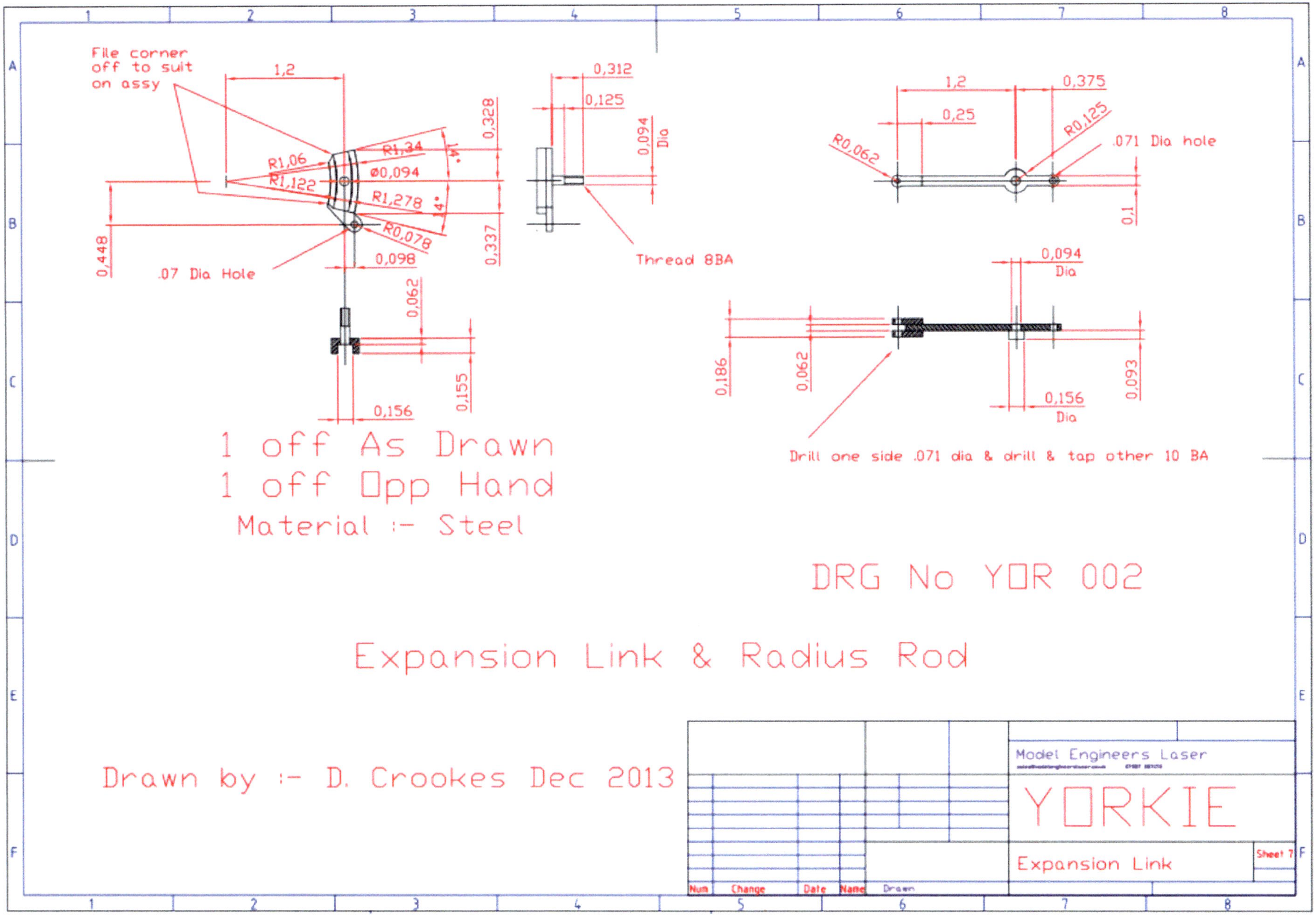

File corner off to suit on assy
1,2
R1,06
R1,122
R1,34
ø0,094
R1,278
R0,078
14°
14°
0,328
0,337
0,448
.07 Dia Hole
0,098
0,062
0,155
0,156
0,312
0,125
0,094 Dia
Thread 8BA
1,2
0,375
0,25
R0,062
R0,125
.071 Dia hole
0,1
0,094 Dia
0,186
0,062
0,156 Dia
0,093
Drill one side .071 dia & drill & tap other 10 BA
1 off As Drawn
1 off Opp Hand
Material :- Steel
DRG No YOR 002
Expansion Link & Radius Rod
Drawn by :- D. Crookes Dec 2013
Model Engineers Laser
YORKIE
Expansion Link
Sheet 7
Num
Change
Date
Name
Drawn

EXPANSION LINK SUPPORT BRACKET

Material :- Steel & brass

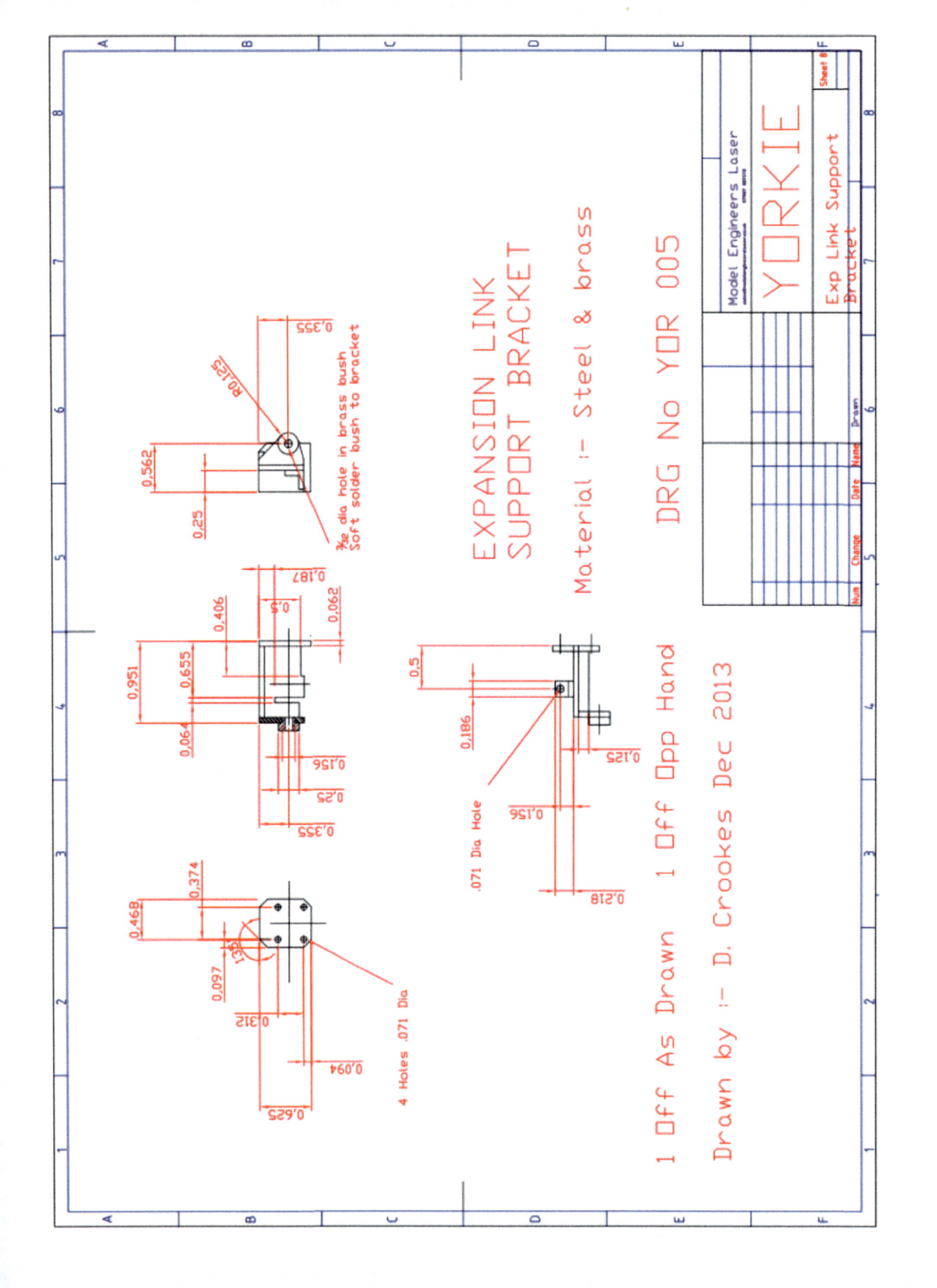

DRG No YOR 005

1 Off As Drawn   1 Off Opp Hand

Drawn by :- D. Crookes Dec 2013

Model Engineers Laser

YORKIE

Exp Link Support Bracket

- 0.355
- 0.562
- 0.25
- R0.125
- 3/32 dia hole in brass bush
- Soft solder bush to bracket
- 0.187
- 0.062
- 0.5
- 0.406
- 0.951
- 0.655
- 0.064
- 0.156
- 0.25
- 0.355
- 0.5
- 0.186
- .071 Dia Hole
- 0.125
- 0.156
- 0.218
- 0.374
- 0.468
- 0.097
- 135°
- 0.312
- 0.094
- 0.625
- 4 Holes .071 Dia

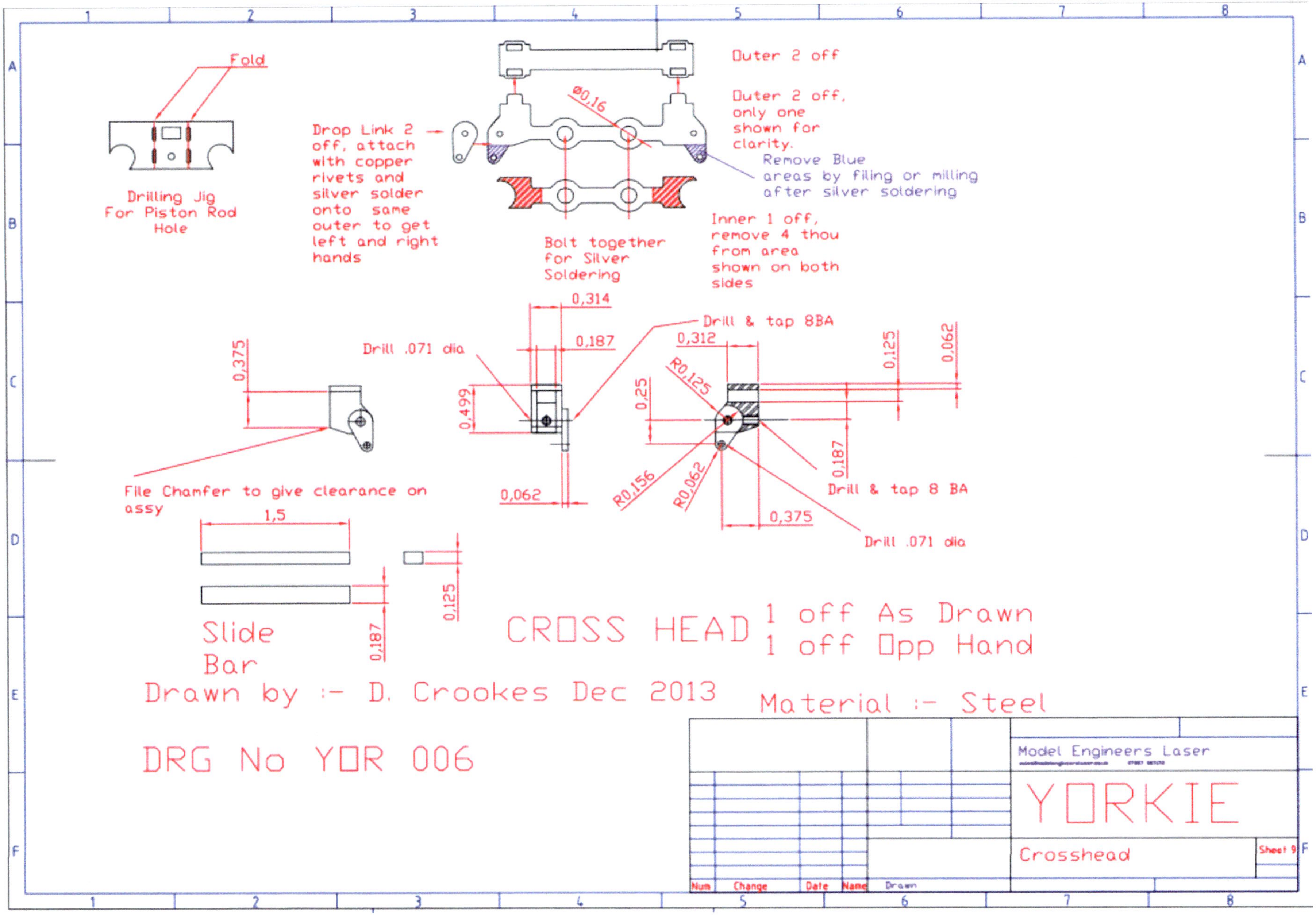

Fold
Drilling Jig For Piston Rod Hole
Drop Link 2 off, attach with copper rivets and silver solder onto same outer to get left and right hands
Ø0,16
Outer 2 off
Outer 2 off, only one shown for clarity.
Remove Blue areas by filing or milling after silver soldering
Bolt together for Silver Soldering
Inner 1 off, remove 4 thou from area shown on both sides
0,314
0,187
Drill .071 dia
0,499
0,062
Drill & tap 8BA
0,312
R0,125
0,25
0,125
0,062
0,187
R0,156
R0,062
Drill & tap 8 BA
0,375
Drill .071 dia
0,375
File Chamfer to give clearance on assy
1,5
0,125
0,187
Slide Bar
CROSS HEAD 1 off As Drawn
1 off Opp Hand
Drawn by :- D. Crookes Dec 2013
Material :- Steel
DRG No YOR 006
Model Engineers Laser
YORKIE
Crosshead
Sheet 9
Num
Change
Date
Name
Drawn

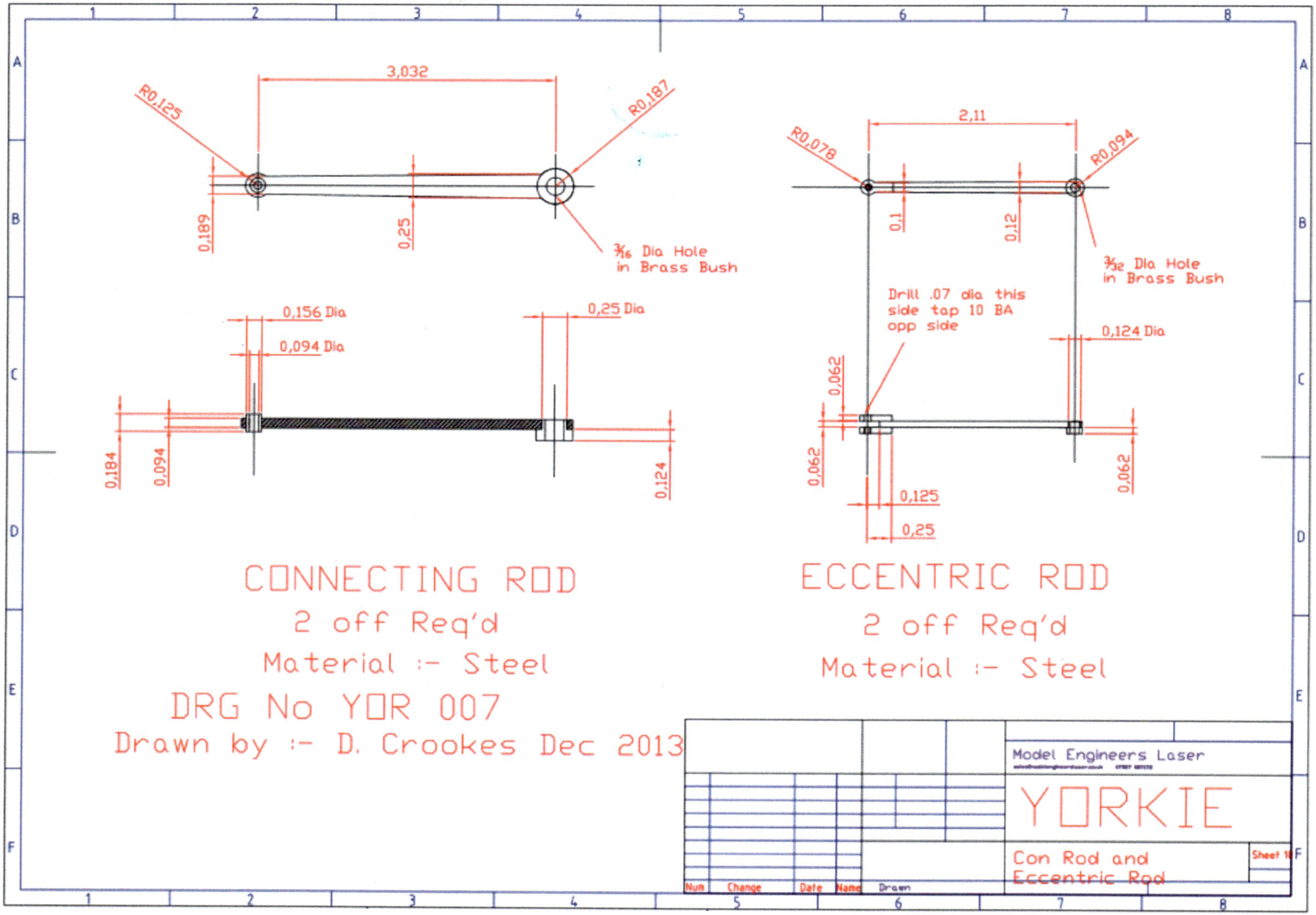
3,032
R0,125
R0,187
0,189
0,25
3⁄16 Dia Hole
in Brass Bush
0,156 Dia
0,094 Dia
0,25 Dia
0,184
0,094
0,124
2,11
R0,078
R0,094
0,1
0,12
3⁄32 Dia Hole
in Brass Bush
Drill .07 dia this
side tap 10 BA
opp side
0,124 Dia
0,062
0,062
0,062
0,125
0,25
CONNECTING ROD
2 off Req'd
Material :- Steel
DRG No YOR 007
Drawn by :- D. Crookes Dec 2013
ECCENTRIC ROD
2 off Req'd
Material :- Steel
Model Engineers Laser
YORKIE
Con Rod and
Eccentric Rod
Sheet 1
Num
Change
Date
Name
Drawn

YORKIE
Coupling rods etc

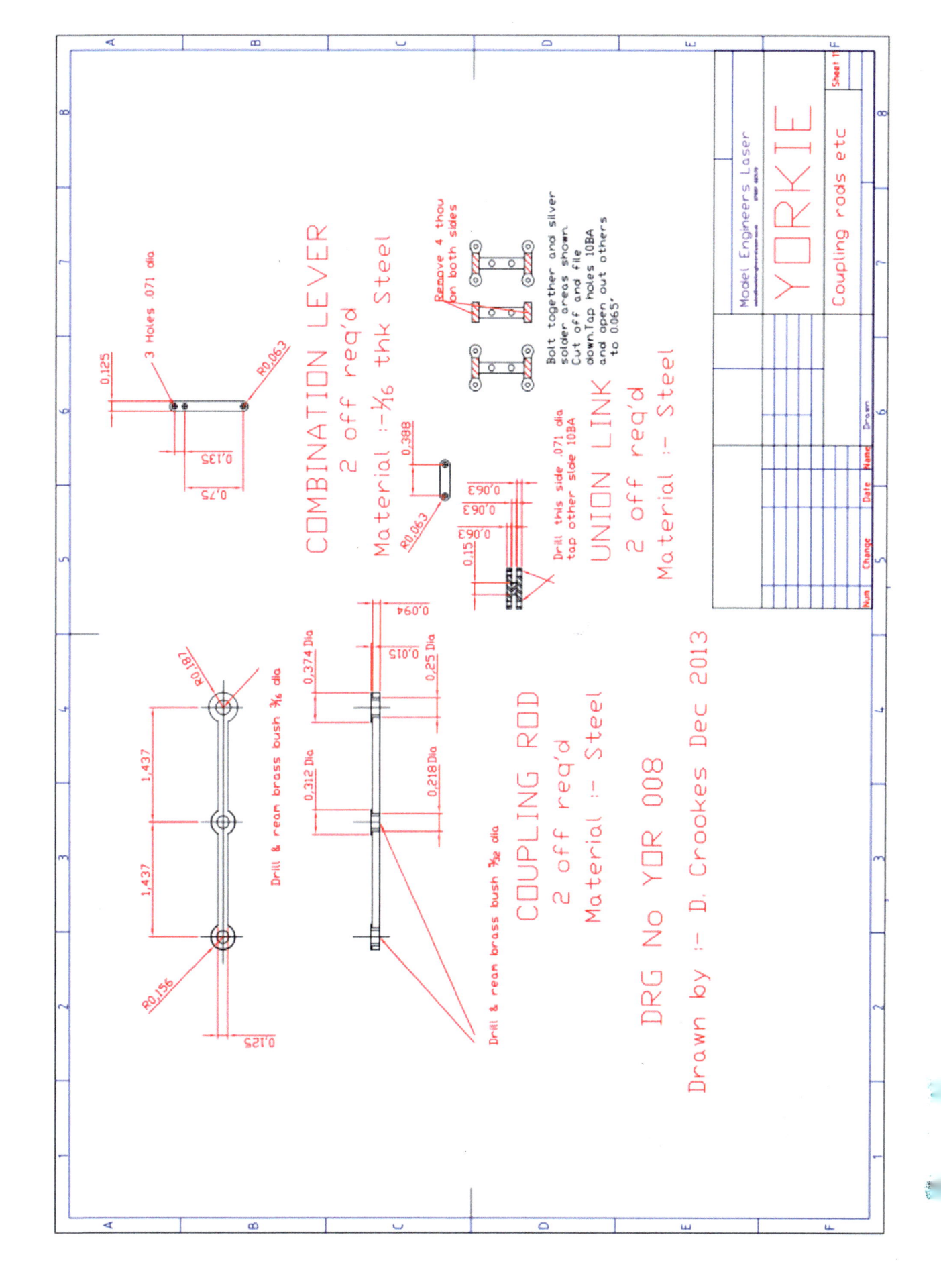

COMBINATION LEVER
2 off req'd
Material :- 3/16 thk Steel

- 3 Holes .071 dia
- R0.063
- 0.125
- 0.135
- 0.75
- 0.388
- R0.063
- 0.15
- 0.063
- 0.063
- 0.063
- 0.063

UNION LINK
2 off req'd
Material :- Steel

Drill this side .071 dia tap other side 10BA

Bolt together and silver solder areas shown. Cut off and file down. Tap holes 10BA and open out others to 0.065"

Remove 4 thou on both sides

COUPLING ROD
2 off req'd
Material :- Steel

- R0.187
- 1.437
- 1.437
- R0.156
- 0.125
- 0.374 Dia
- 0.094
- 0.015
- 0.25 Dia
- 0.312 Dia
- 0.218 Dia

Drill & ream brass bush 3/16 dia
Drill & ream brass bush 3/32 dia

DRG No YOR 008
Drawn by :- D. Crookes Dec 2013

Model Engineers Laser

YORKIE

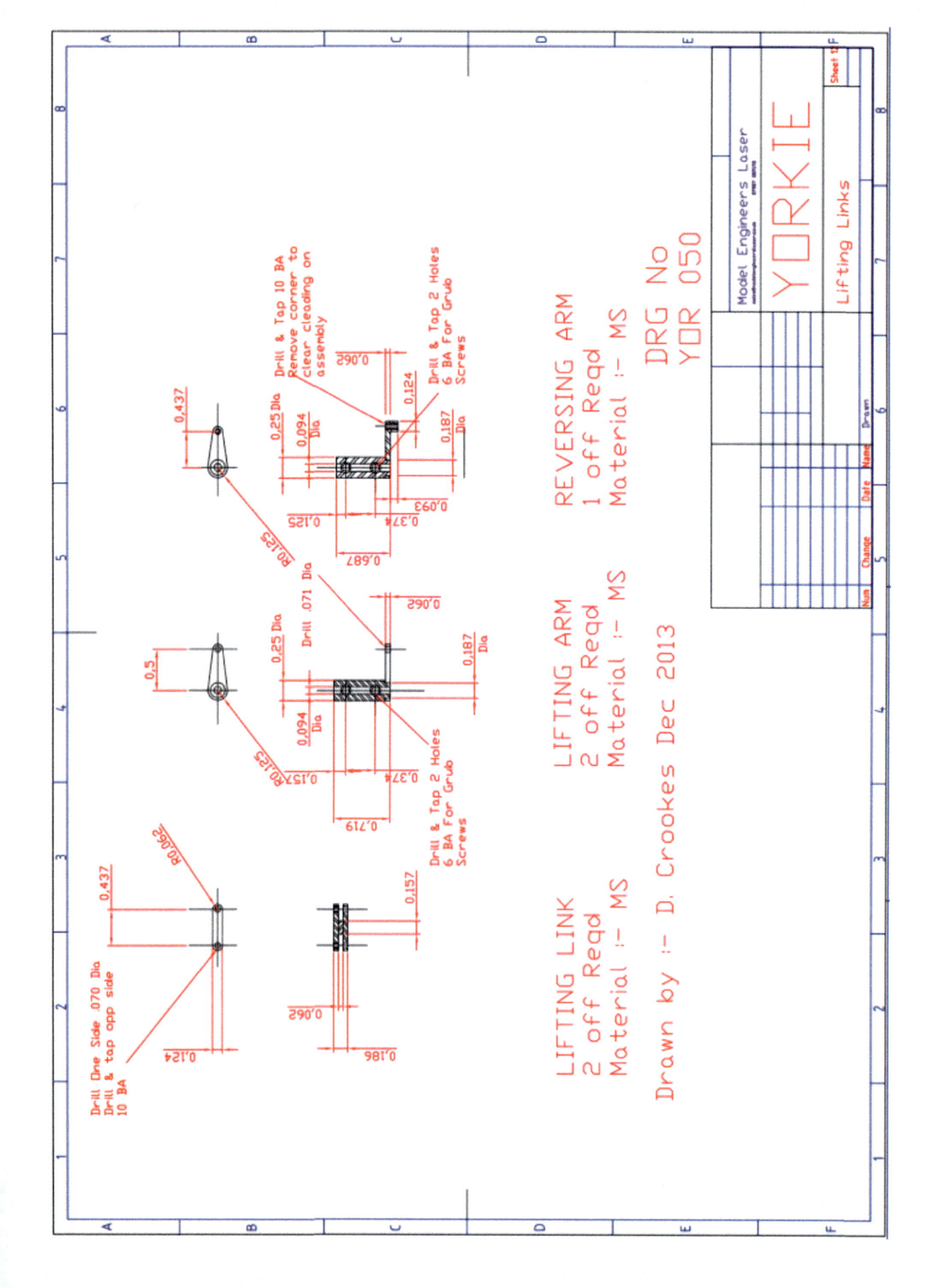

Lifting Links

DRG No YOR 050

LIFTING LINK
2 off Reqd
Material :- MS

Drill One Side .070 Dia
Drill & tap opp side
10 BA

0.437
R0.062
0.124
0.062
0.186
0.157

LIFTING ARM
2 off Reqd
Material :- MS

0.5
0.25 Dia
Drill .071 Dia
0.094 Dia
R0.125
0.062
0.187 Dia
0.37
0.719
0.157

Drill & Tap 2 Holes
6 BA For Grub Screws

REVERSING ARM
1 off Reqd
Material :- MS

0.437
0.25 Dia
0.094 Dia
R0.125
0.062
0.124
0.187 Dia
0.37
0.093
0.687

Drill & Tap 10 BA
Remove corner to clear cleading on assembly

Drill & Tap 2 Holes
6 BA For Grub Screws

Drawn by :- D. Crookes Dec 2013

Model Engineers Laser

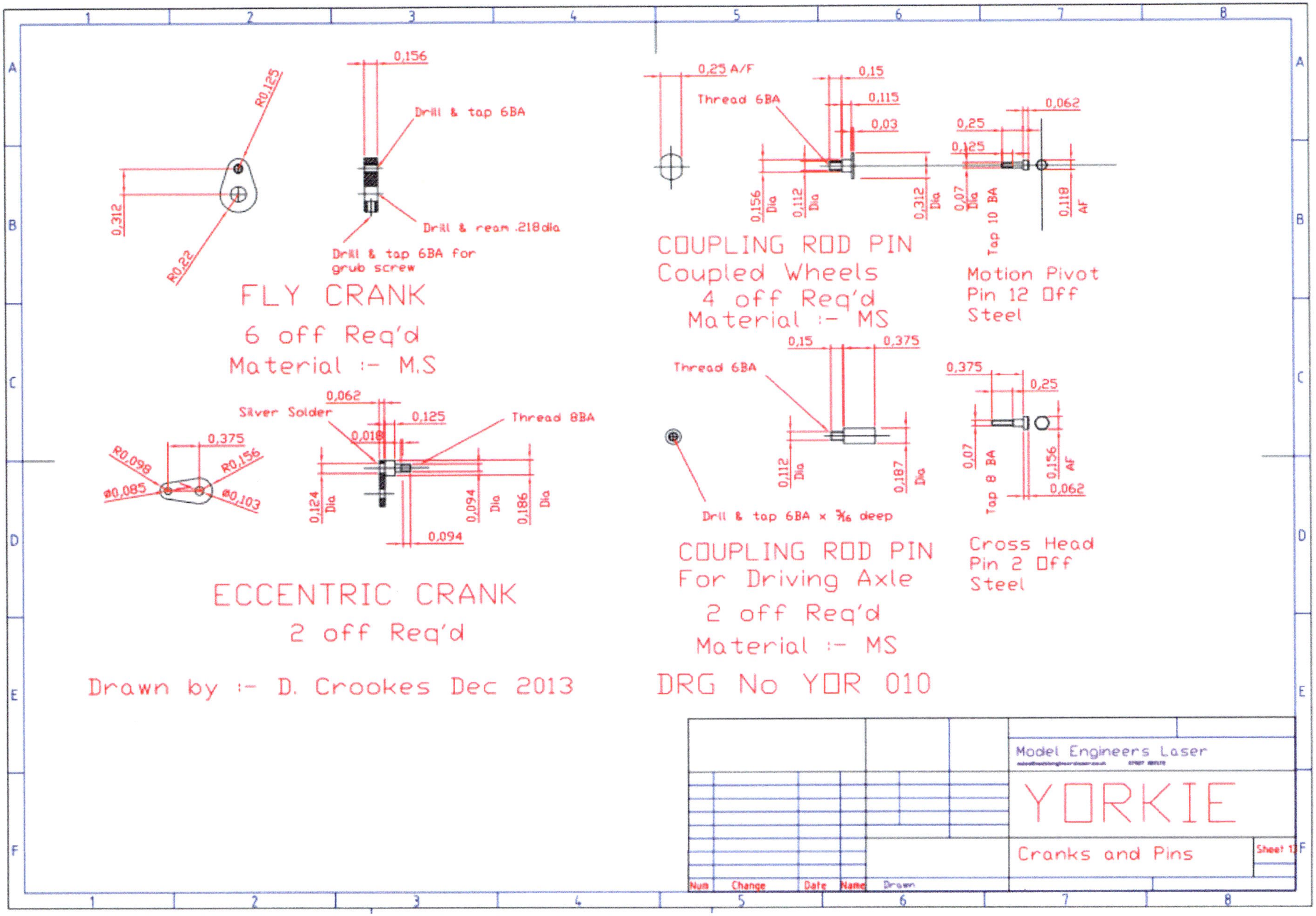

R0,125
R0,22
0,312
0,156
Drill & tap 6BA
Drill & ream .218dia
Drill & tap 6BA for grub screw
FLY CRANK
6 off Req'd
Material :- M.S
R0,098
Ø0,085
0,375
R0,156
Ø0,103
0,062
Silver Solder
0,018
0,125
Thread 8BA
0,124 Dia
0,094 Dia
0,186 Dia
0,094
ECCENTRIC CRANK
2 off Req'd
Drawn by :- D. Crookes Dec 2013
0,25 A/F
Thread 6BA
0,15
0,115
0,03
0,156 Dia
0,112 Dia
0,312 Dia
COUPLING ROD PIN
Coupled Wheels
4 off Req'd
Material :- MS
0,25
0,125
0,062
0,07 Dia
Tap 10 BA
0,118 AF
Motion Pivot Pin 12 Off Steel
0,15
0,375
Thread 6BA
0,112 Dia
0,187 Dia
Drill & tap 6BA x 3⁄16 deep
COUPLING ROD PIN
For Driving Axle
2 off Req'd
Material :- MS
0,375
0,25
0,07
Tap 8 BA
0,156 AF
0,062
Cross Head Pin 2 Off Steel
DRG No YOR 010
Model Engineers Laser
YORKIE
Cranks and Pins
Num
Change
Date
Name
Drawn

YORKIE
Wheels and axles

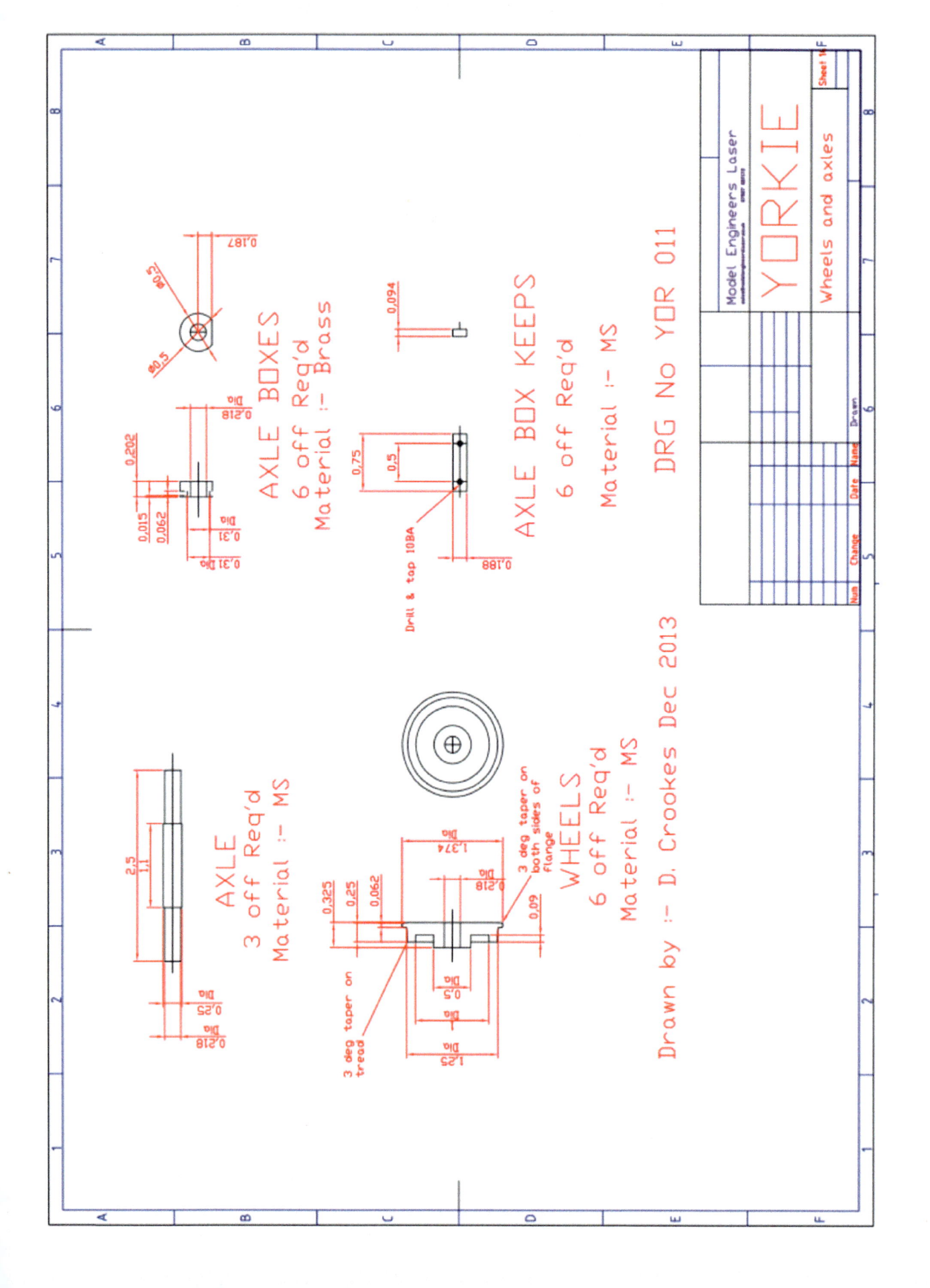

AXLE BOXES
6 off Req'd
Material :- Brass

AXLE BOX KEEPS
6 off Req'd
Material :- MS

Drill & tap 10BA

AXLE
3 off Req'd
Material :- MS

WHEELS
6 off Req'd
Material :- MS

3 deg taper on tread
3 deg taper on both sides of flange

Drawn by :- D. Crookes Dec 2013

DRG No YOR 011

Model Engineers Laser

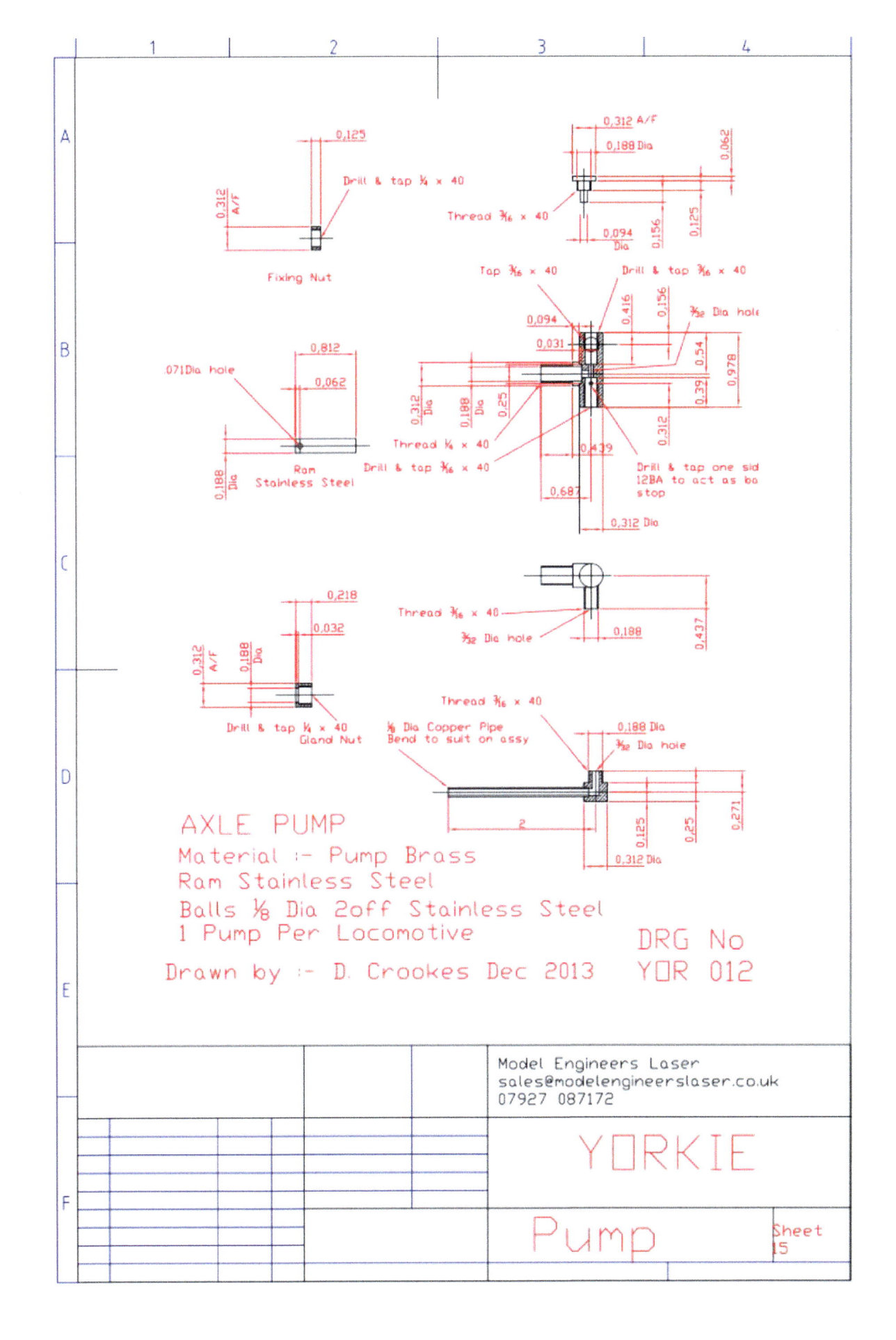

AXLE PUMP
Material :- Pump Brass
Ram Stainless Steel
Balls ⅛ Dia 2off Stainless Steel
1 Pump Per Locomotive

Drawn by :- D. Crookes Dec 2013

DRG No YOR 012

Model Engineers Laser
sales@modelengineerslaser.co.uk
07927 087172

YORKIE

Pump

Sheet 15

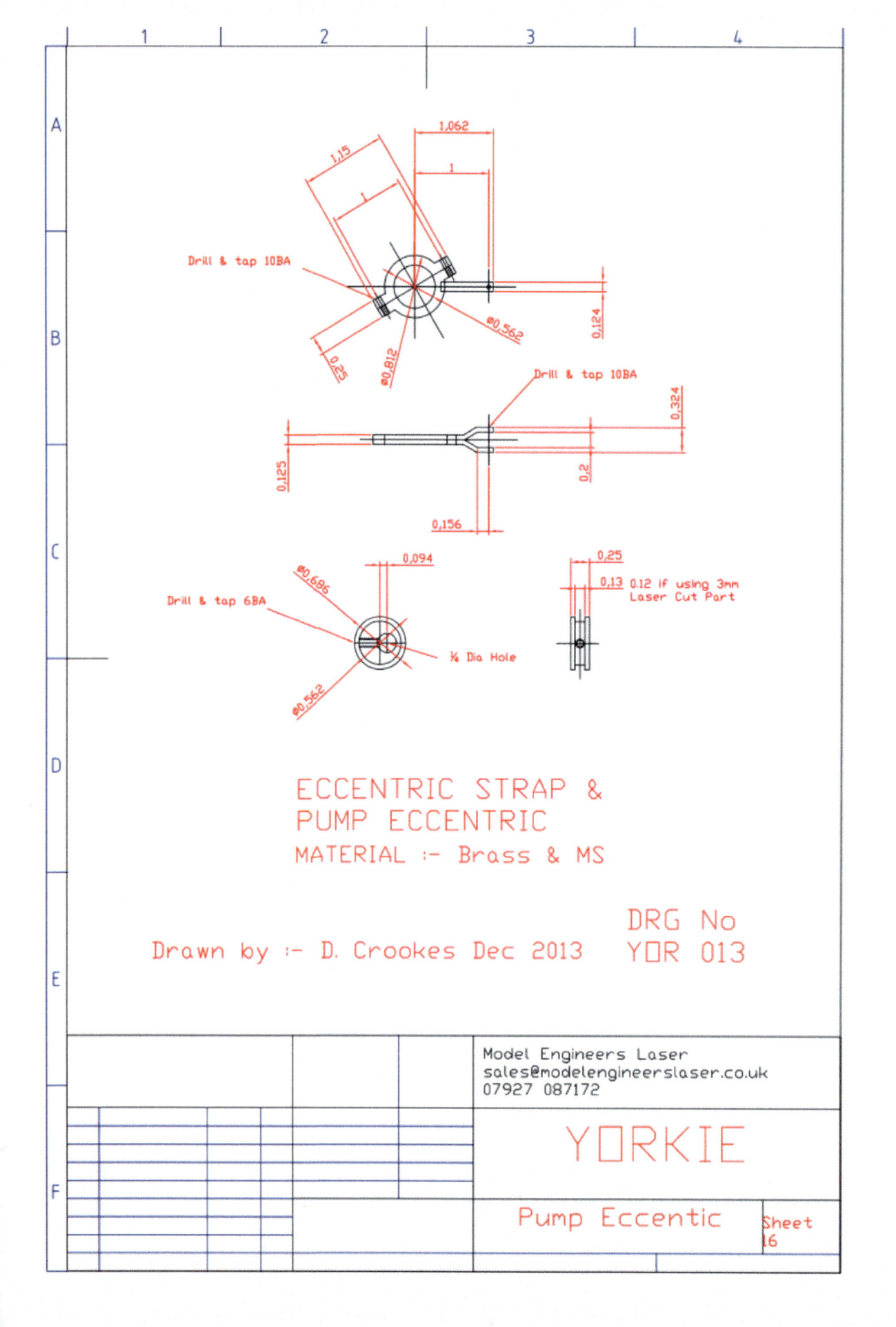

ECCENTRIC STRAP & PUMP ECCENTRIC
MATERIAL :- Brass & MS

Drawn by :- D. Crookes Dec 2013

DRG No YOR 013

Model Engineers Laser
sales@modelengineerslaser.co.uk
07927 087172

YORKIE

Pump Eccentic

Sheet 16

YORKIE Smokebox

SMOKEBOX

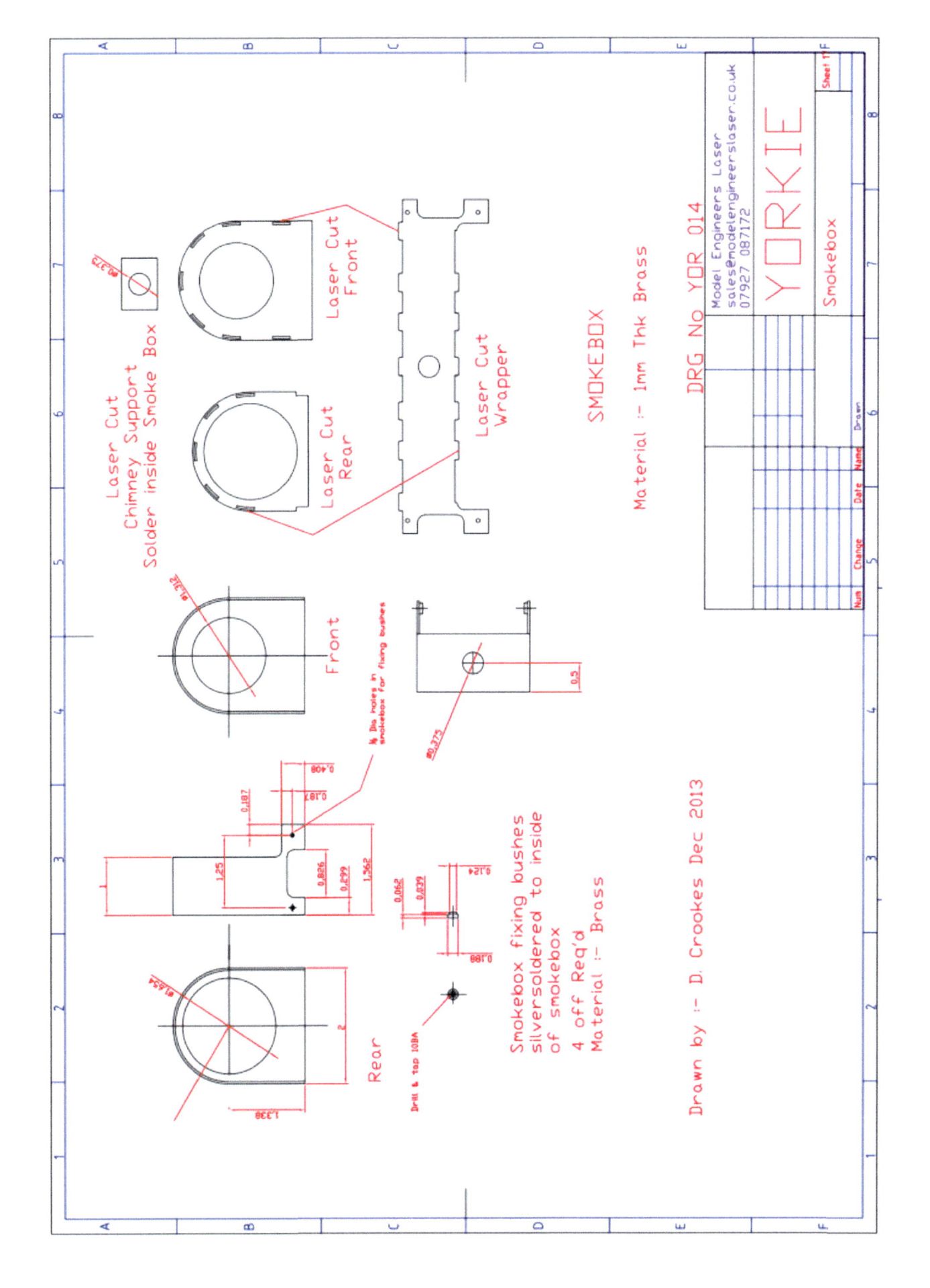

Material :- 1mm Thk Brass

DRG No YOR 014

Laser Cut Front
Laser Cut Rear
Laser Cut Wrapper
Laser Cut Chimney Support — Solder inside Smoke Box

¼ Dia holes in smokebox for fixing bushes

Smokebox fixing bushes silversoldered to inside of smokebox
4 off Req'd
Material :- Brass

Drill & tap 10BA

Drawn by :- D. Crookes Dec 2013

Model Engineers Laser
sales@modelengineerslaser.co.uk
07927 087172

CHIMNEY
Material :- MS

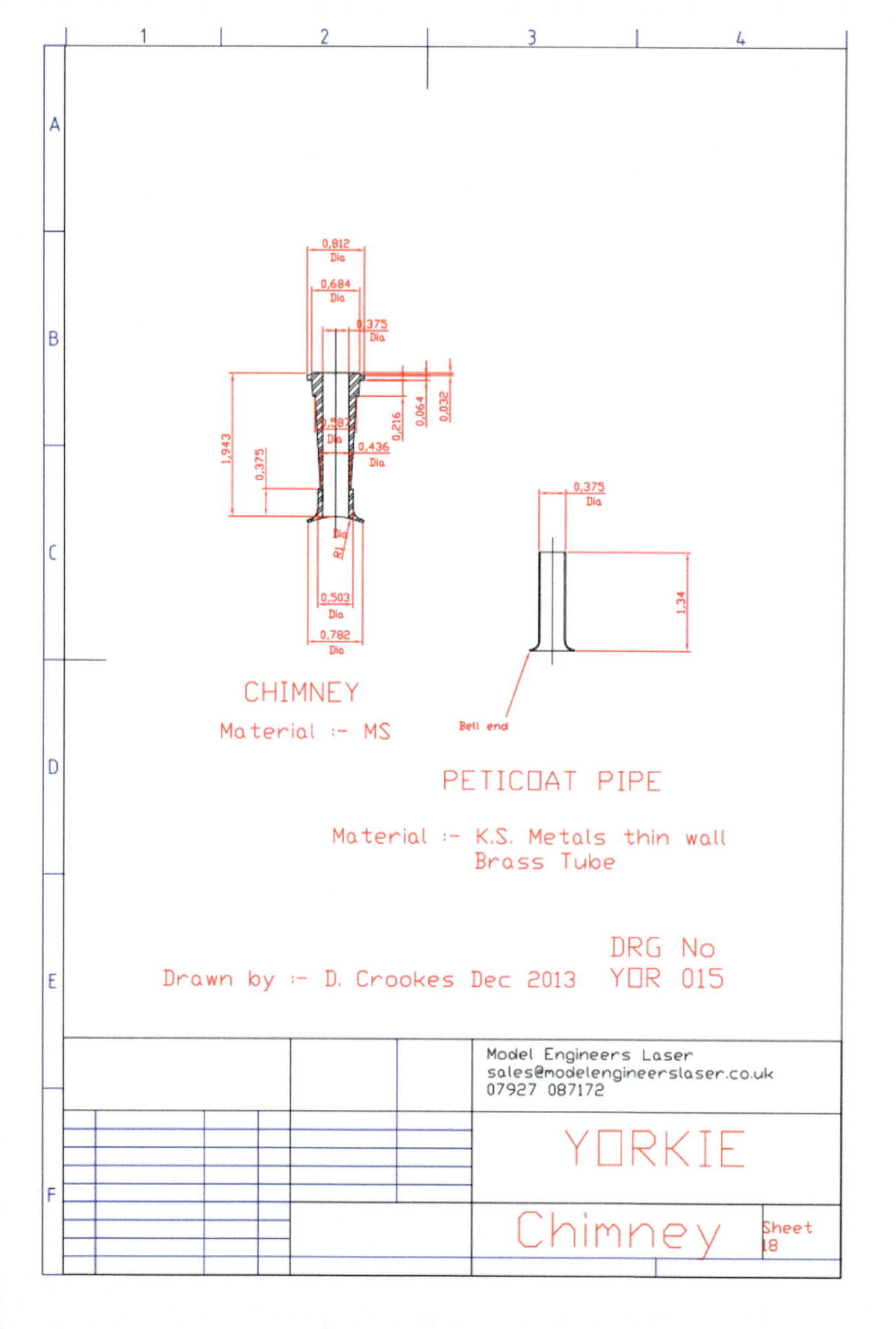

PETICOAT PIPE
Material :- K.S. Metals thin wall Brass Tube

Bell end

Drawn by :- D. Crookes Dec 2013

DRG No YOR 015

Model Engineers Laser
sales@modelengineerslaser.co.uk
07927 087172

YORKIE

Chimney

Sheet 18

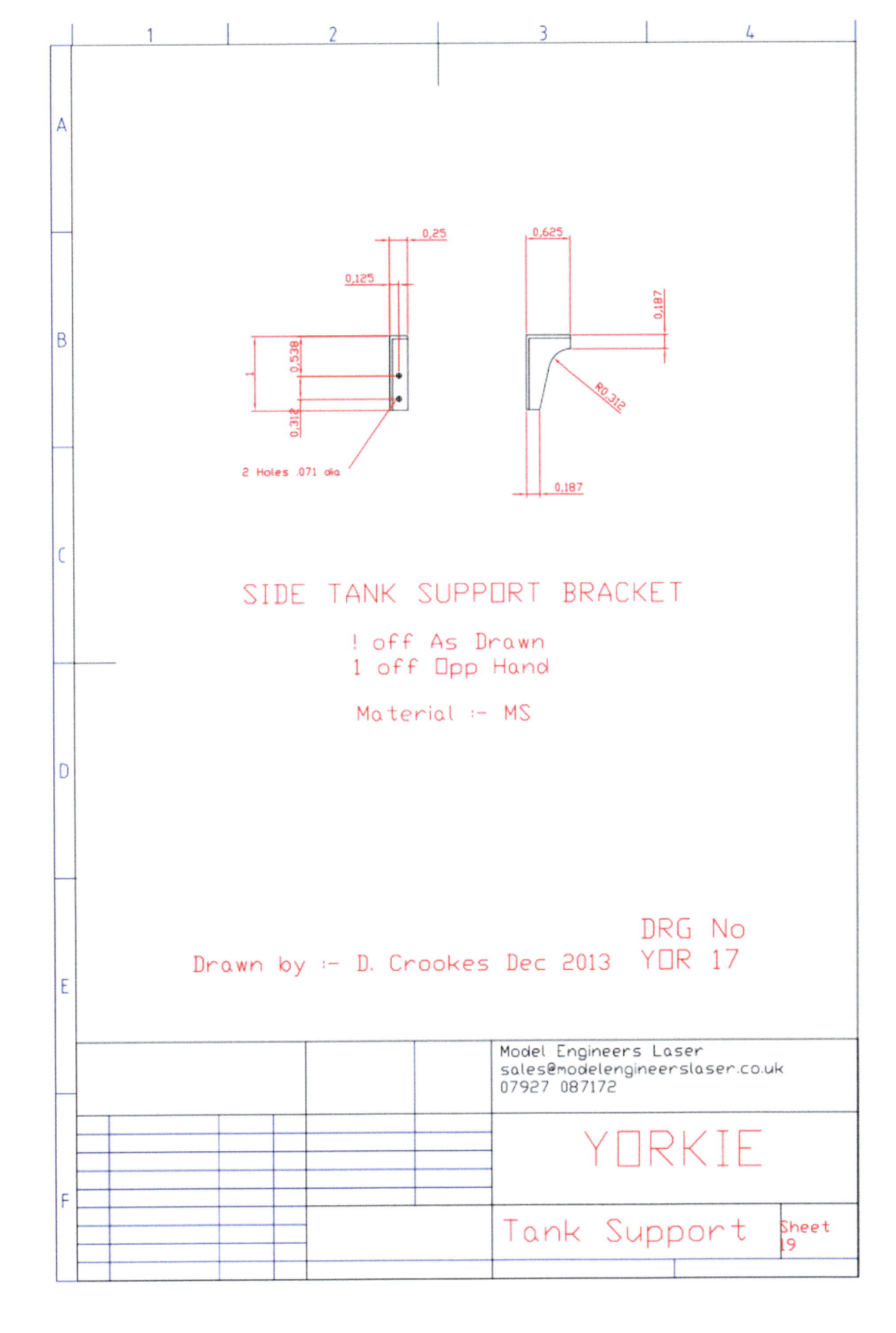

SIDE TANK SUPPORT BRACKET

1 off As Drawn
1 off Opp Hand

Material :- MS

2 Holes .071 dia

Drawn by :- D. Crookes Dec 2013   DRG No YOR 17

Model Engineers Laser
sales@modelengineerslaser.co.uk
07927 087172

YORKIE

Tank Support   Sheet 19

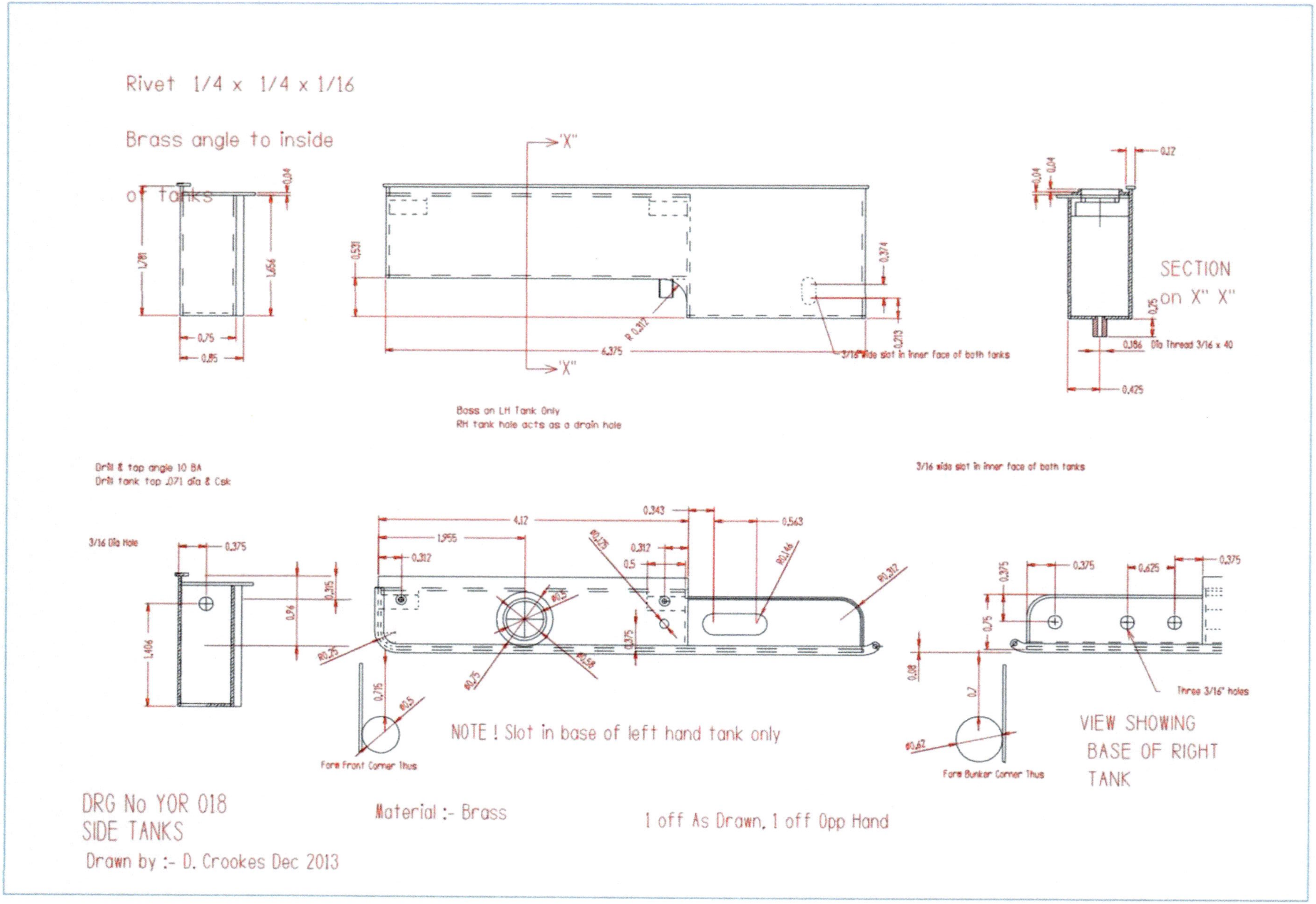

Rivet 1/4 x 1/4 x 1/16
Brass angle to inside
of tanks
0.04
1.781
1.656
0.75
0.85
"X"
"X"
0.531
R 0.312
6.375
0.374
0.213
3/16 wide slot in inner face of both tanks
0.12
0.04
0.04
SECTION
on X" X"
0.25
0.186
Dia Thread 3/16 x 40
0.425
Boss on LH Tank Only
RH tank hole acts as a drain hole
Drill & tap angle 10 BA
Drill tank top .071 dia & Csk
3/16 wide slot in inner face of both tanks
3/16 Dia Hole
0.375
1.406
0.96
0.315
4.12
1.955
0.312
0.343
0.563
0.312
0.5
Ø0.125
R0.146
R0.312
Ø0.5
R0.25
Ø0.58
Ø0.75
0.375
0.715
Ø0.5
Form Front Corner Thus
NOTE ! Slot in base of left hand tank only
0.375
0.375
0.625
0.375
0.75
0.08
0.7
Ø0.62
Three 3/16" holes
Form Bunker Corner Thus
VIEW SHOWING
BASE OF RIGHT
TANK
DRG No YOR 018
SIDE TANKS
Drawn by :- D. Crookes Dec 2013
Material :- Brass
1 off As Drawn, 1 off Opp Hand

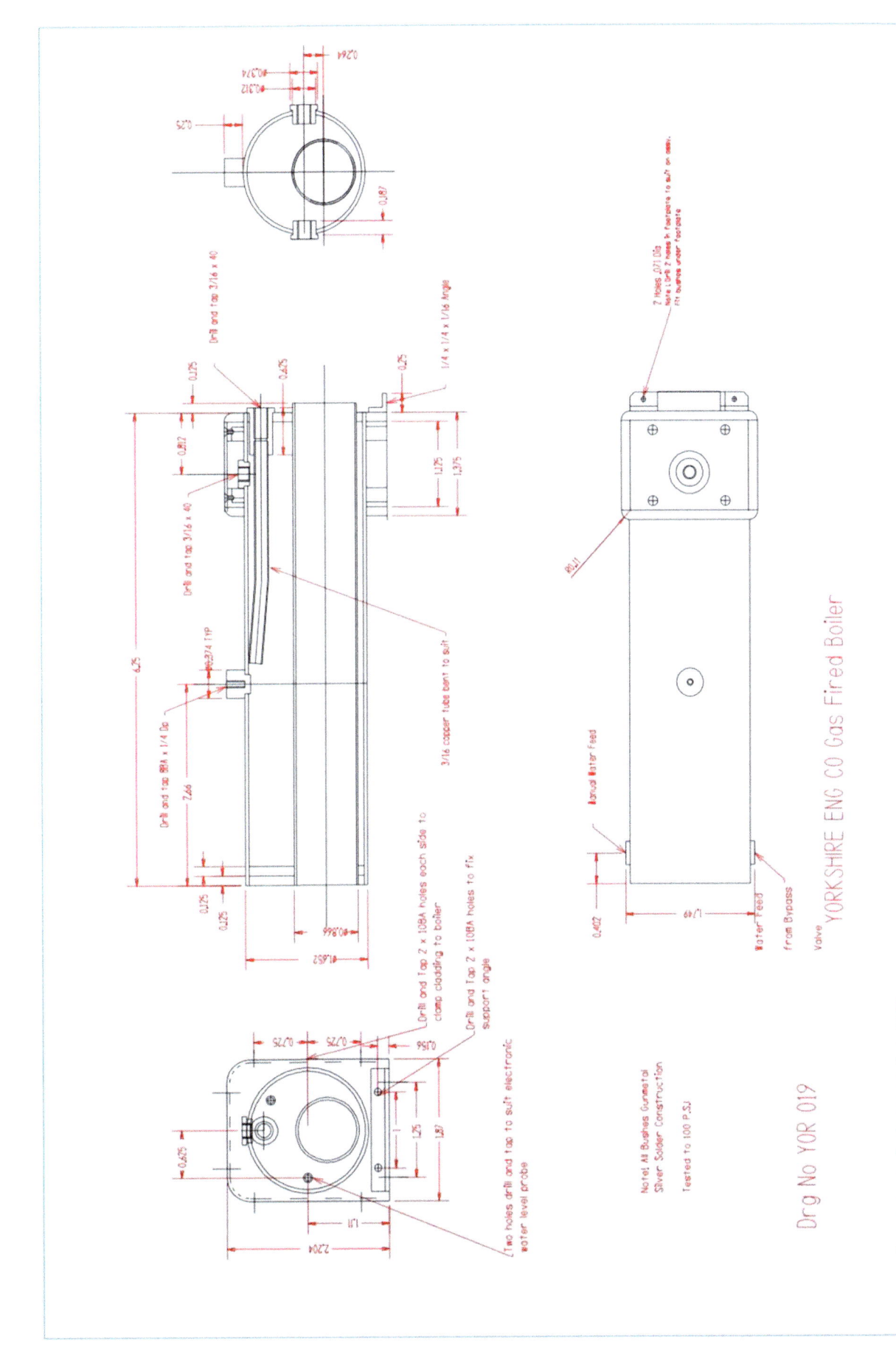

Note!
Use ⅛ Dia St St Ball

Use Lee St St Spring

Ref No CI-010B-02-S

Fit thin locknut after setting valve to 40 PSI

0.024
0.075
Thread 7/32 × 40
0.218 Dia
0.07 Dia
0.05
0.25
0.094 Dia

0.064 Dia
0.085
0.5
0.125
0.125 Dia
0.15 Dia

0.25 A/F
Tap 7/32 × 40

Tap 7/32 × 1/4
0.312 Dia
0.188 Dia
0.28
0.032
0.312
.28 Dia LOCKNUT
5/16 Hex
0.062
0.094 Dia
0.187
Thread 3/16 × 40
0.188 Dia

SAFETY VALVE
Material :- Brass

Drawn by :- D. Crookes Dec 2013
DRG No YOR 020

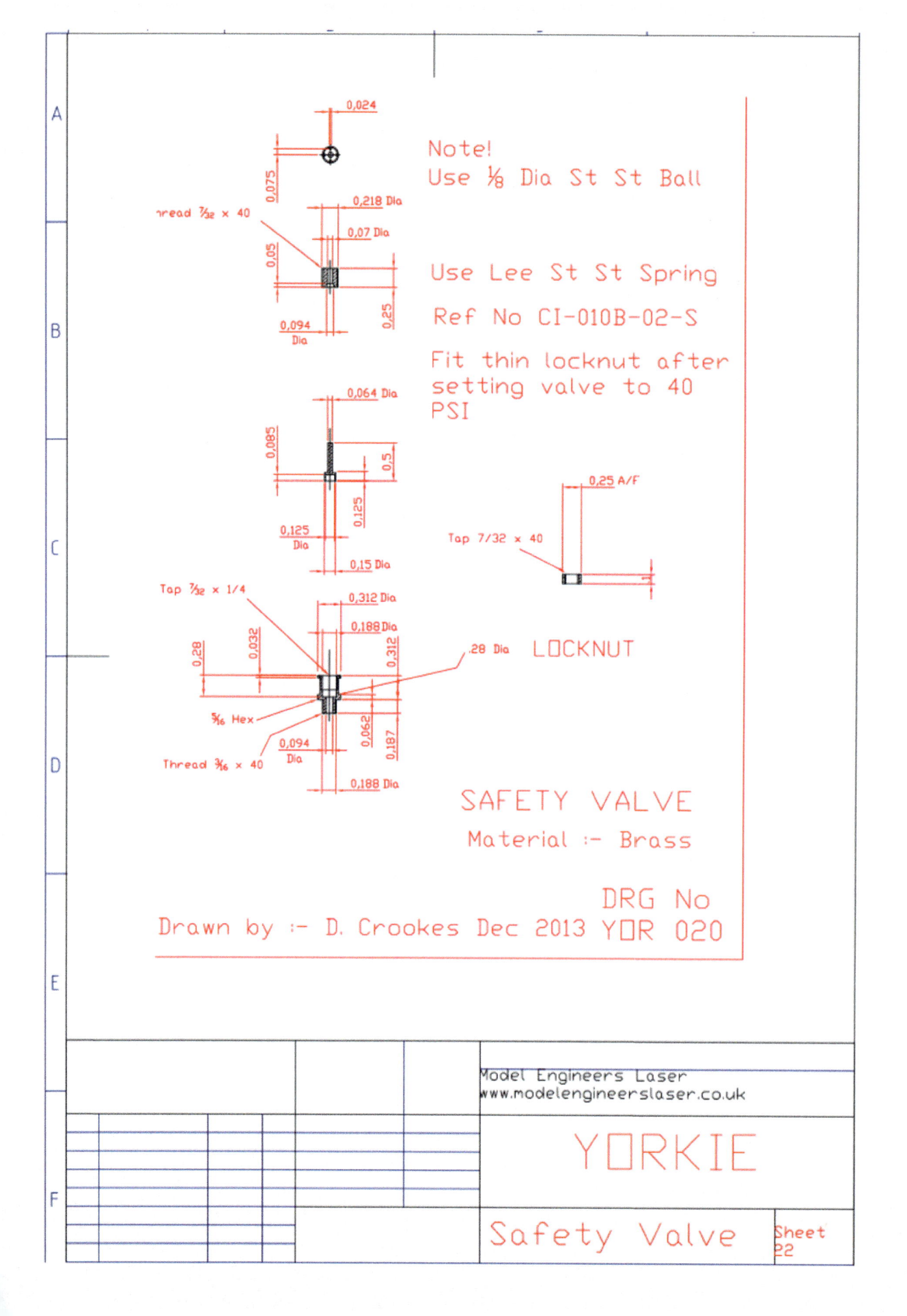

Model Engineers Laser
www.modelengineerslaser.co.uk

YORKIE

Safety Valve

Sheet 22

SAFETY VALVE EXHAUST PIPE

Material :- Copper & Brass

Drawn by :- D. Crookes Dec 2013

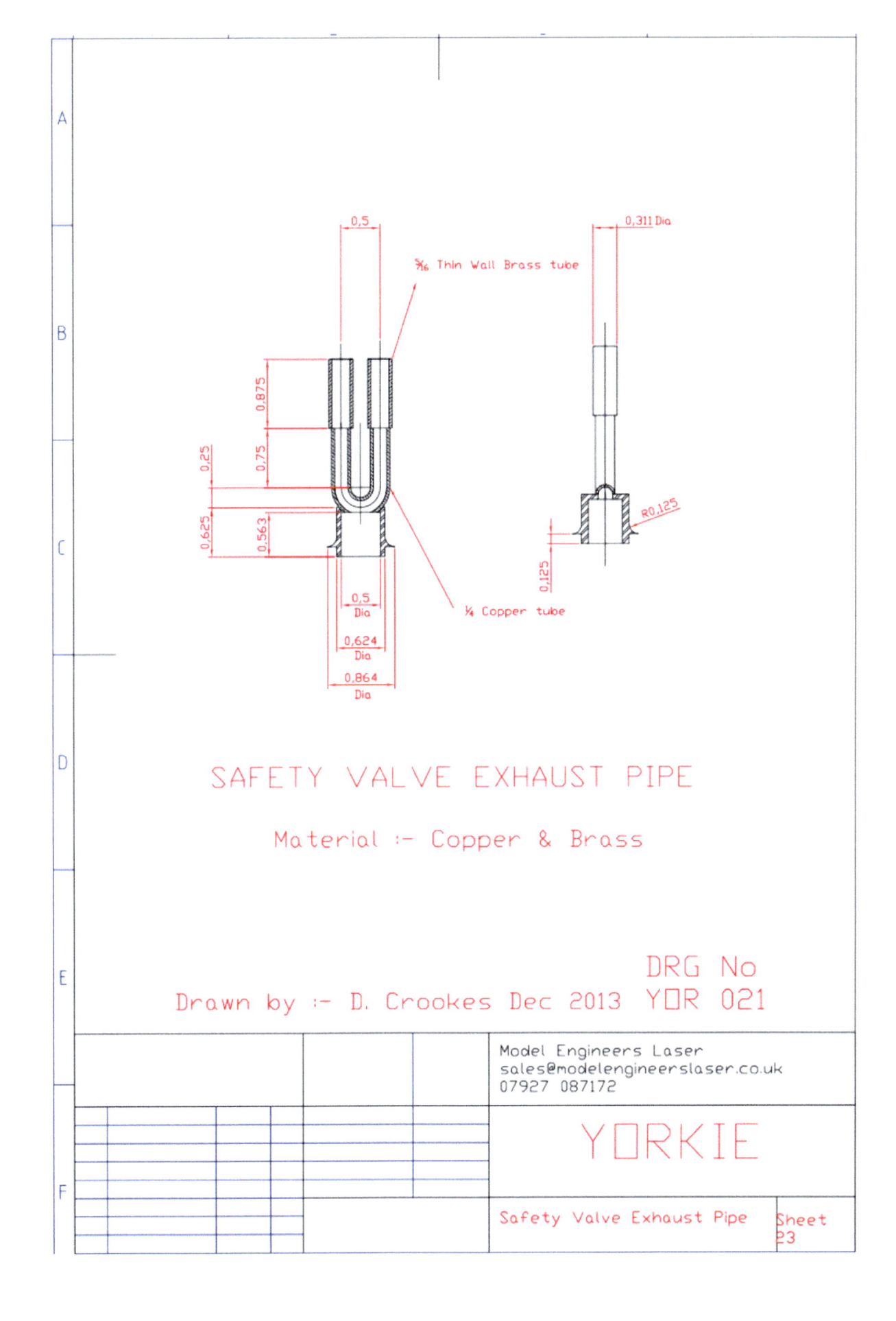

DRG No YOR 021

Model Engineers Laser
sales@modelengineerslaser.co.uk
07927 087172

YORKIE

Safety Valve Exhaust Pipe Sheet 23

BYPASS VALVE

Material :- Brass

LOCKNUT

3mm St ST Screwed Rod
Cone end @ 30 deg

Typical 3/32" OD Pipe End, Make from brass ring solderd onto pipe

Typical Nut Brass

Drawn by :- D. Crookes Dec 2013

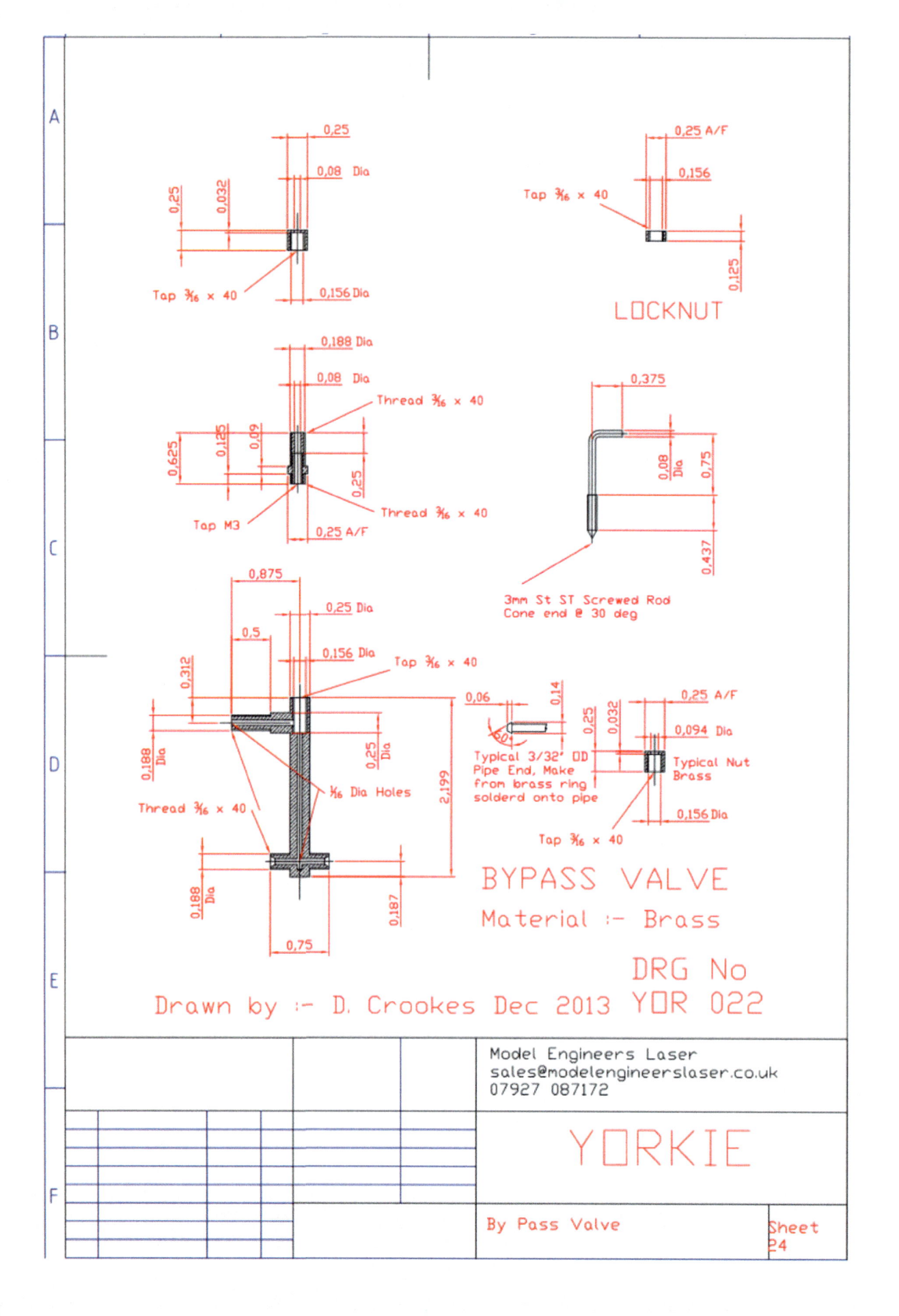

DRG No YOR 022

Model Engineers Laser
sales@modelengineerslaser.co.uk
07927 087172

YORKIE

By Pass Valve

Sheet 24

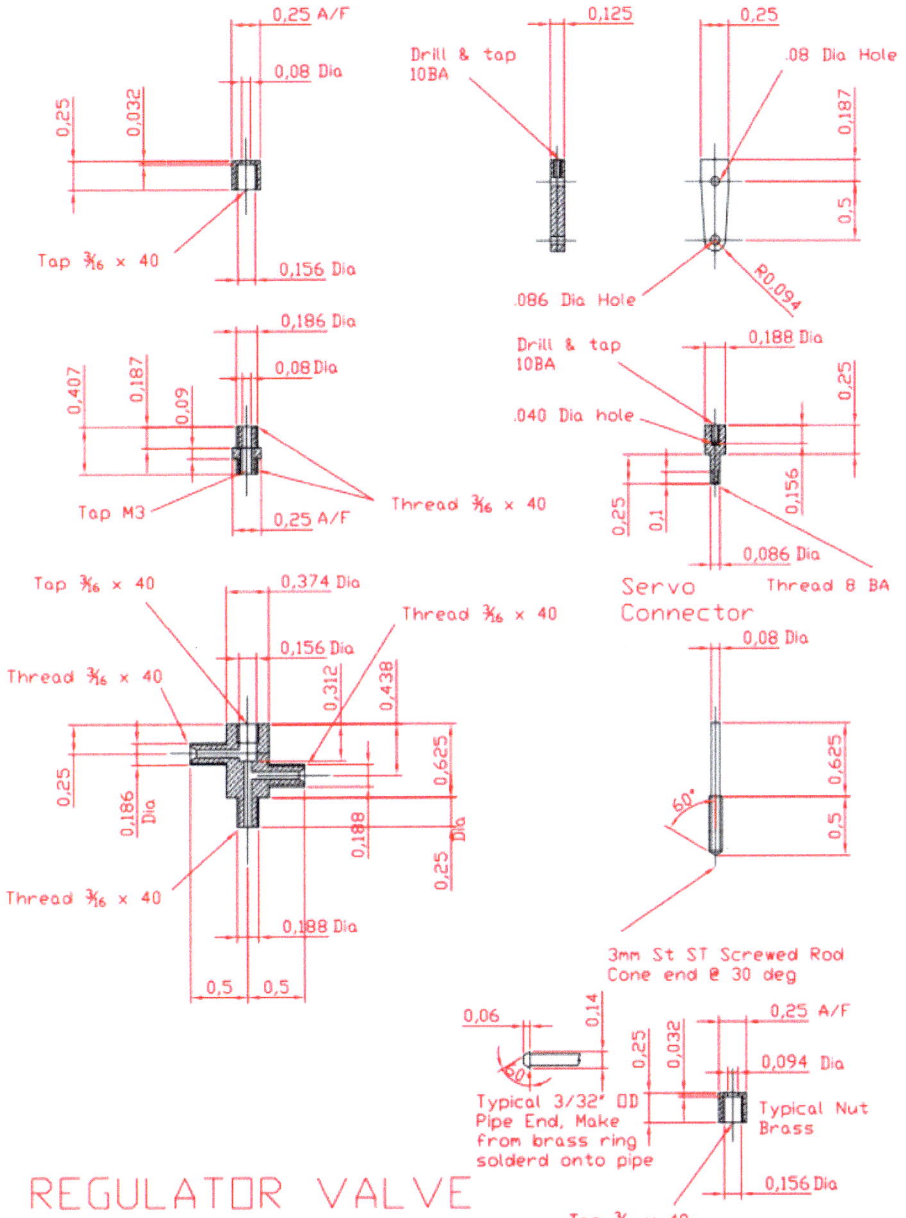

REGULATOR VALVE

Material :- Brass & St St

DRG No

Drawn by :- D. Crookes Dec 2013 YDR 023

Model Engineers Laser
sales@modelengineerslaser.co.uk
07927 087172

YORKIE

Regulator Valve

Sheet
25

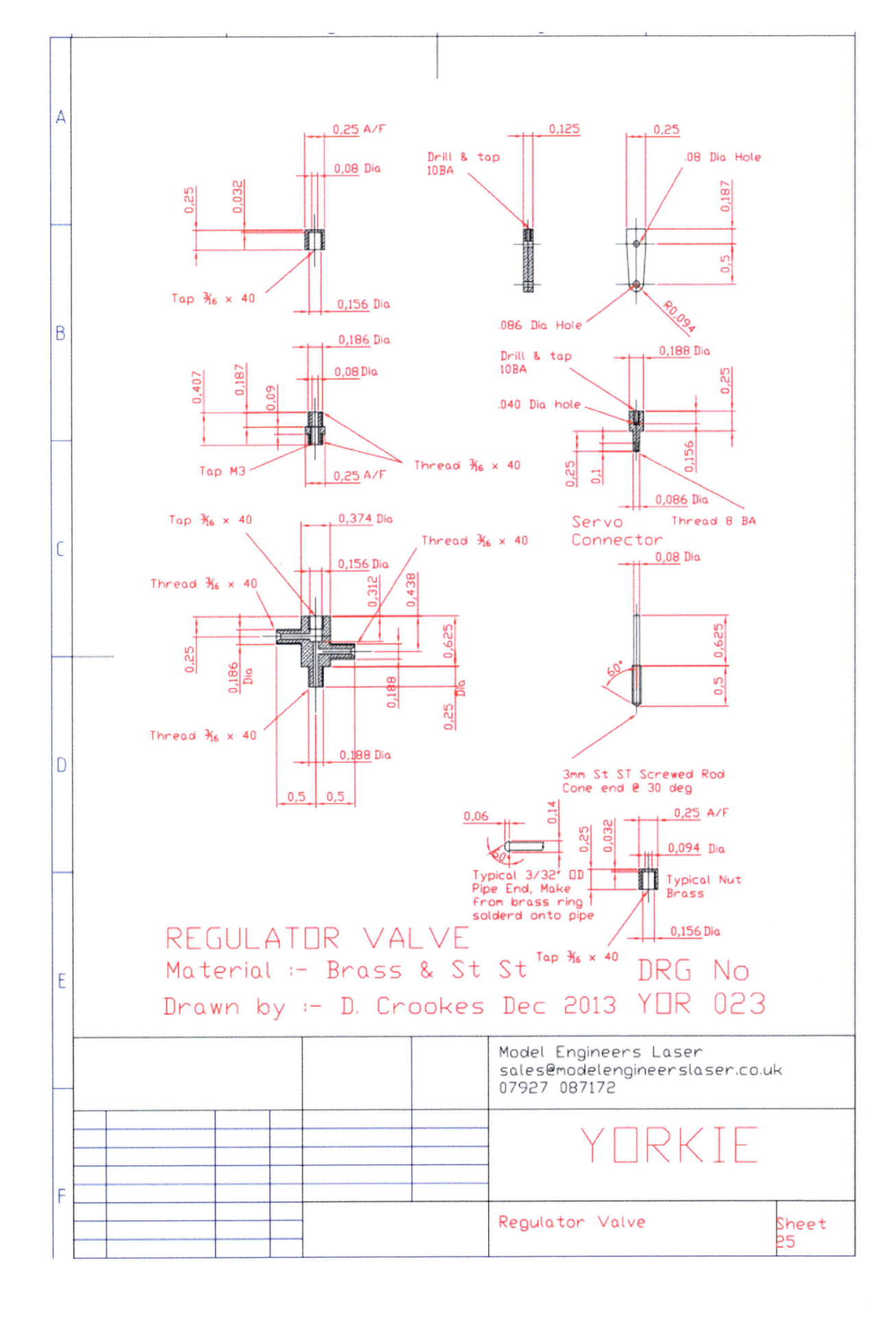

REACH ROD LATCH & REGULATOR SERVO BRACKET

Material :- M.S.

Drill & tap 10BA

Drill & tap tap holes 10BA

.070 Dia holes

Reach Rod Bracket and support for under footplate. Silver solder assembly, file off excess. Fit to footplate and mark holes in frame for bracket. Drill and tap 10BA

Drawn by :- D. Crookes Dec 2013

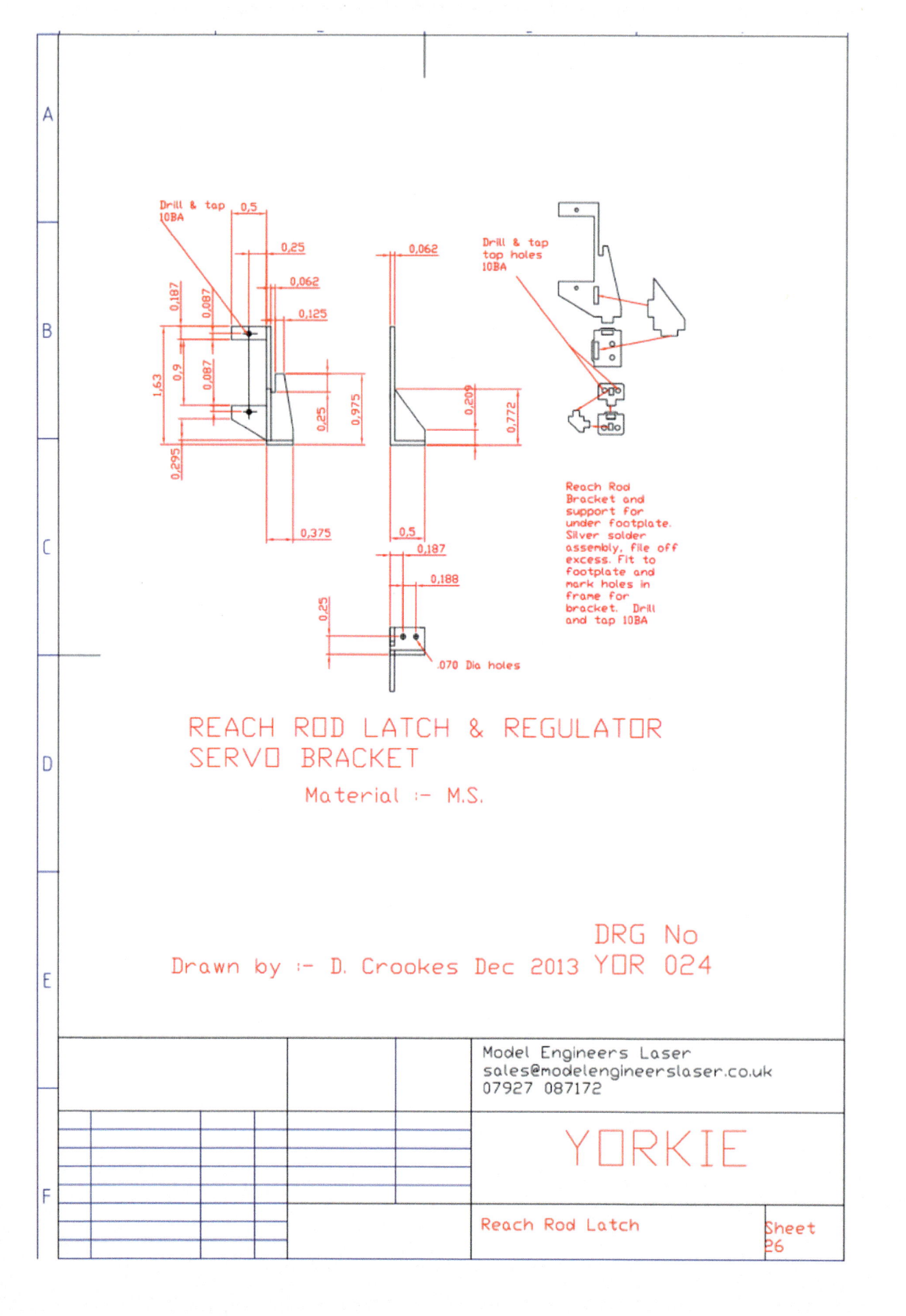

DRG No YOR 024

Model Engineers Laser
sales@modelengineerslaser.co.uk
07927 087172

YORKIE

Reach Rod Latch

Sheet 26

64°

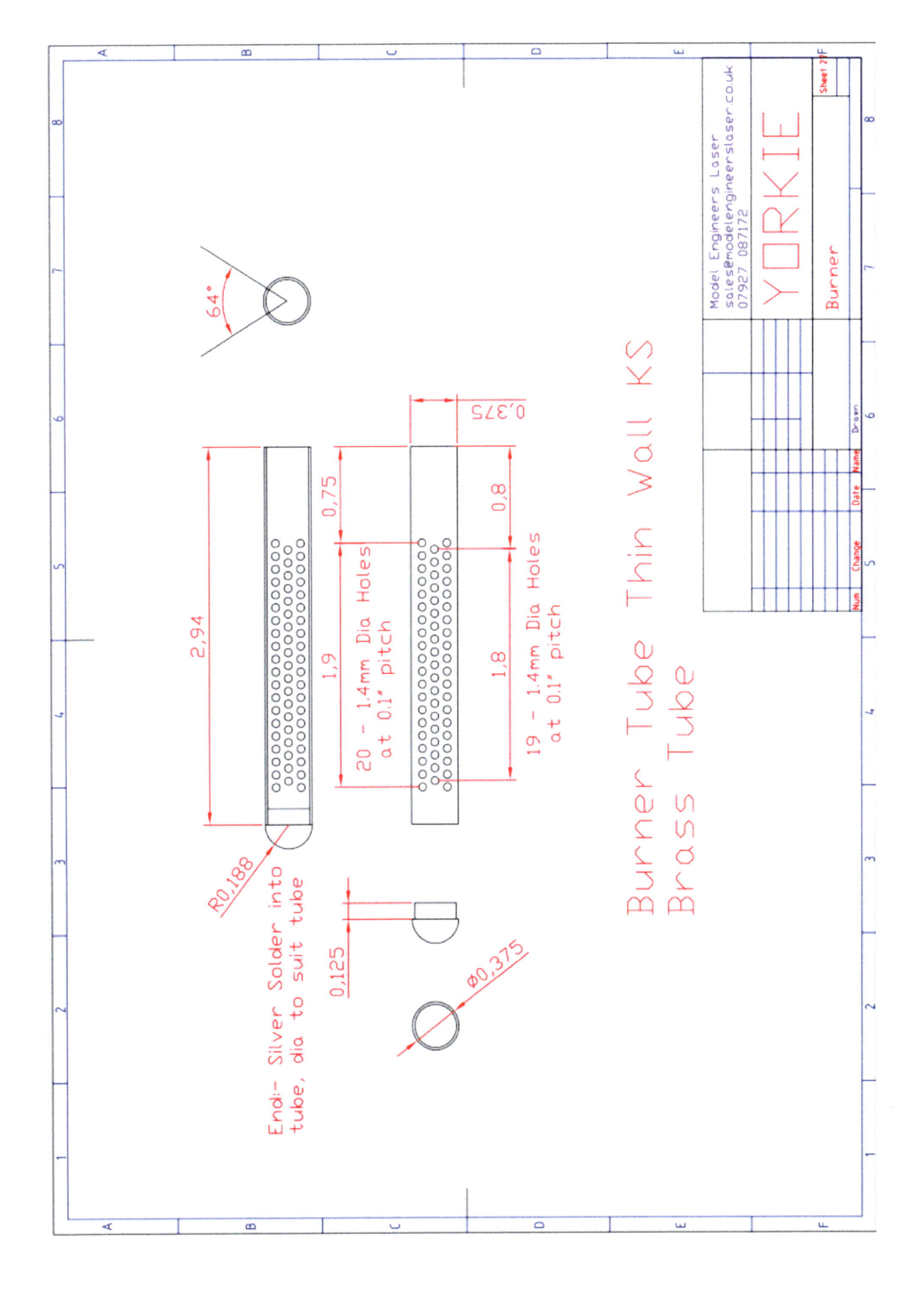

0.375
0.75
0.8
2.94
1.9
1.8

20 - 1.4mm Dia Holes at 0.1" pitch
19 - 1.4mm Dia Holes at 0.1" pitch

R0.188
End:- Silver Solder into tube, dia to suit tube

0.125
Ø0.375

Burner Tube Thin Wall KS
Brass Tube

Model Engineers Laser
sales@modelengineerslaser.co.uk
07927 087172

YORKIE

Burner

Sheet 27F

JET HOLDER
Material :- Brass

BURNER HOLDER
Material :- Brass

NOTE:- JET MUST have PTFE tape in the threads to seal it

NOTE:- This dia to be a press fit into boiler flue tube ⌀0.8

Drill & Tap 1 BA To suit No 5 Gas Jet

³⁄₃₂ Dia hole

⅛ Dia Hole

Drill & tap 10BA

¼" Diameter Hole

Drawn by :- D. Crookes Dec 2013

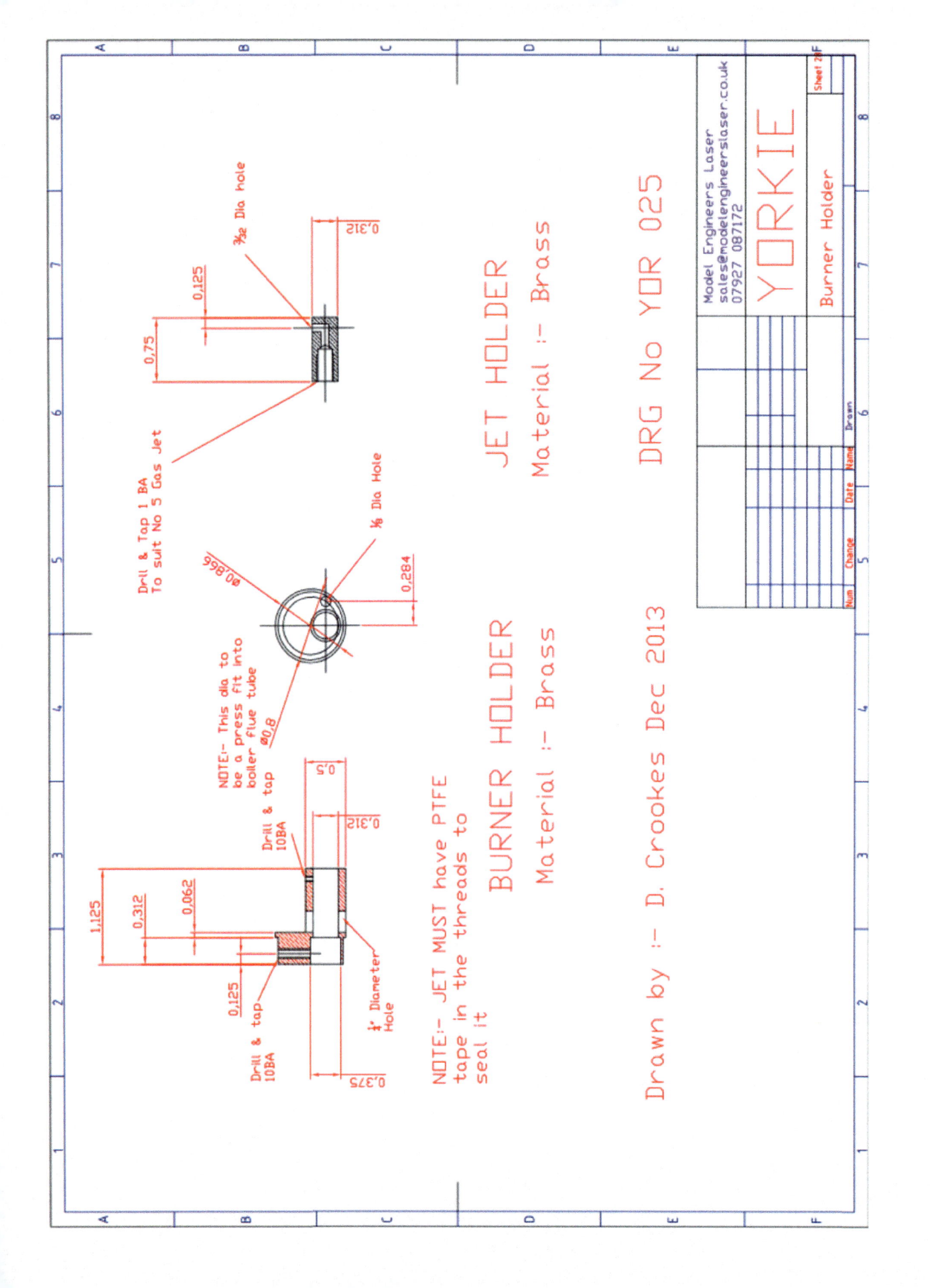

YORKIE

Model Engineers Laser
sales@modelengineerslaser.co.uk
07927 087172

Burner Holder

DRG No YOR 025

Sheet 2F

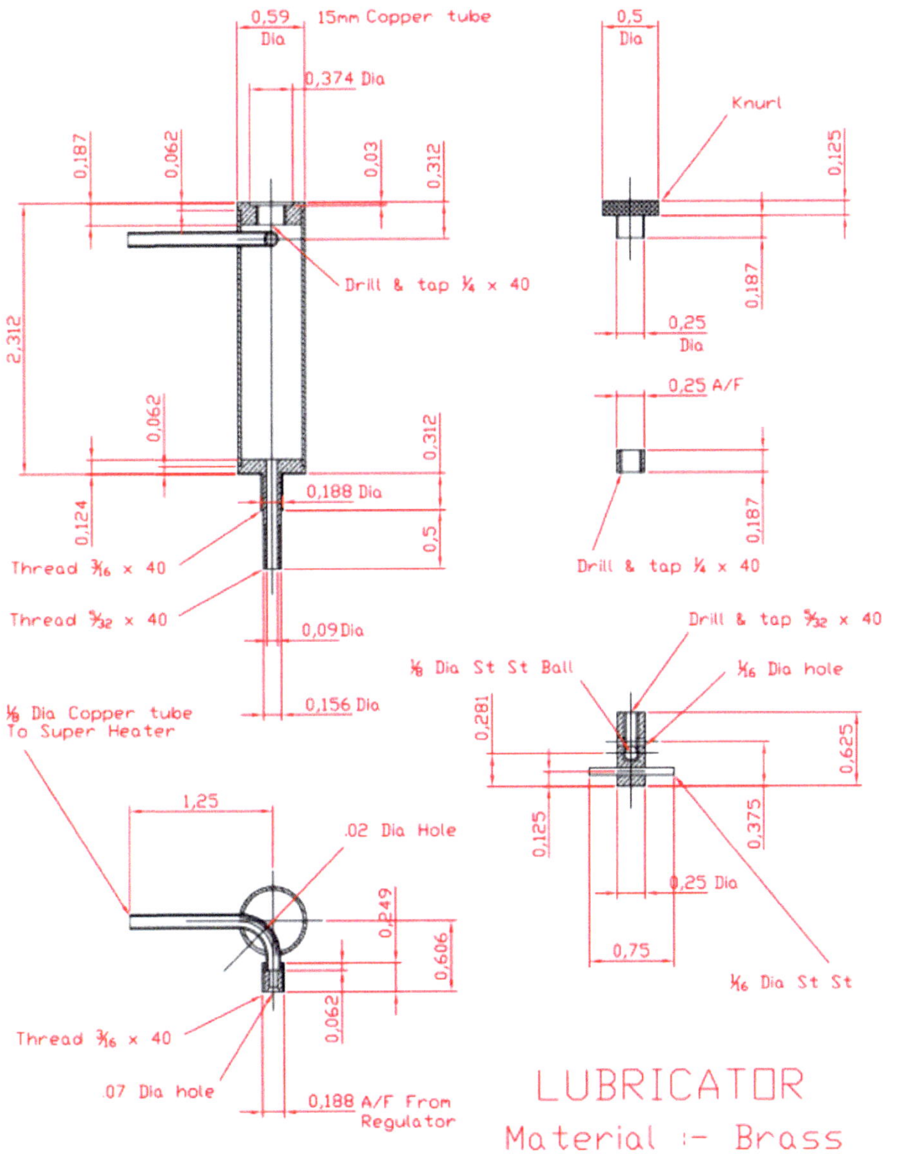

LUBRICATOR

Material :- Brass

DRG No
YOR 026

Drawn by :- D. Crookes Dec 2013

Model Engineers Laser
sales@modelengineerslaser.co.uk
07927 087172

YORKIE

Lubricator

Sheet 29

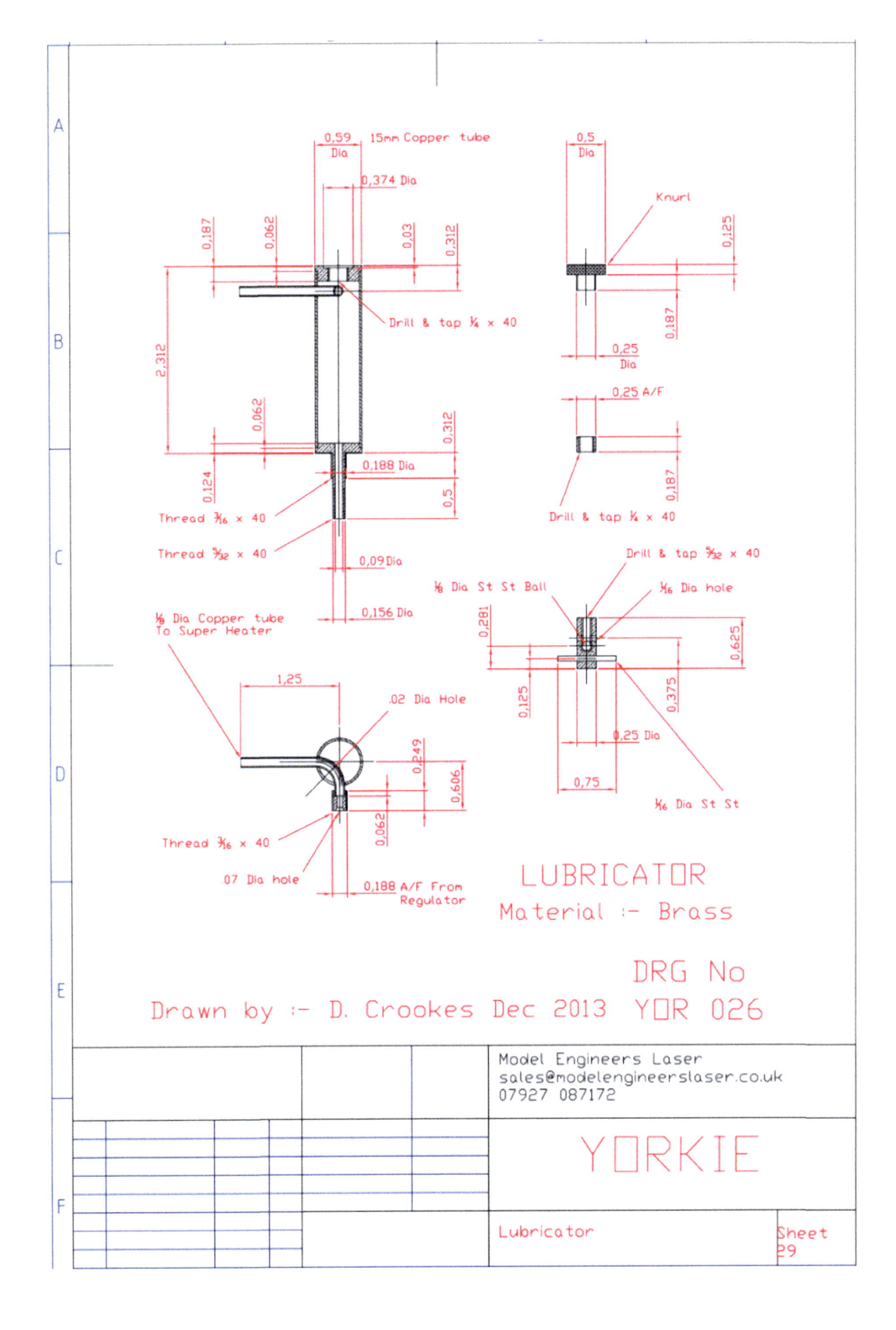

SECTION on X-X

DRG NO YOR 027

CROSS SECTION THRO PUMP

Drawn by :- D. Crookes Dec 2013

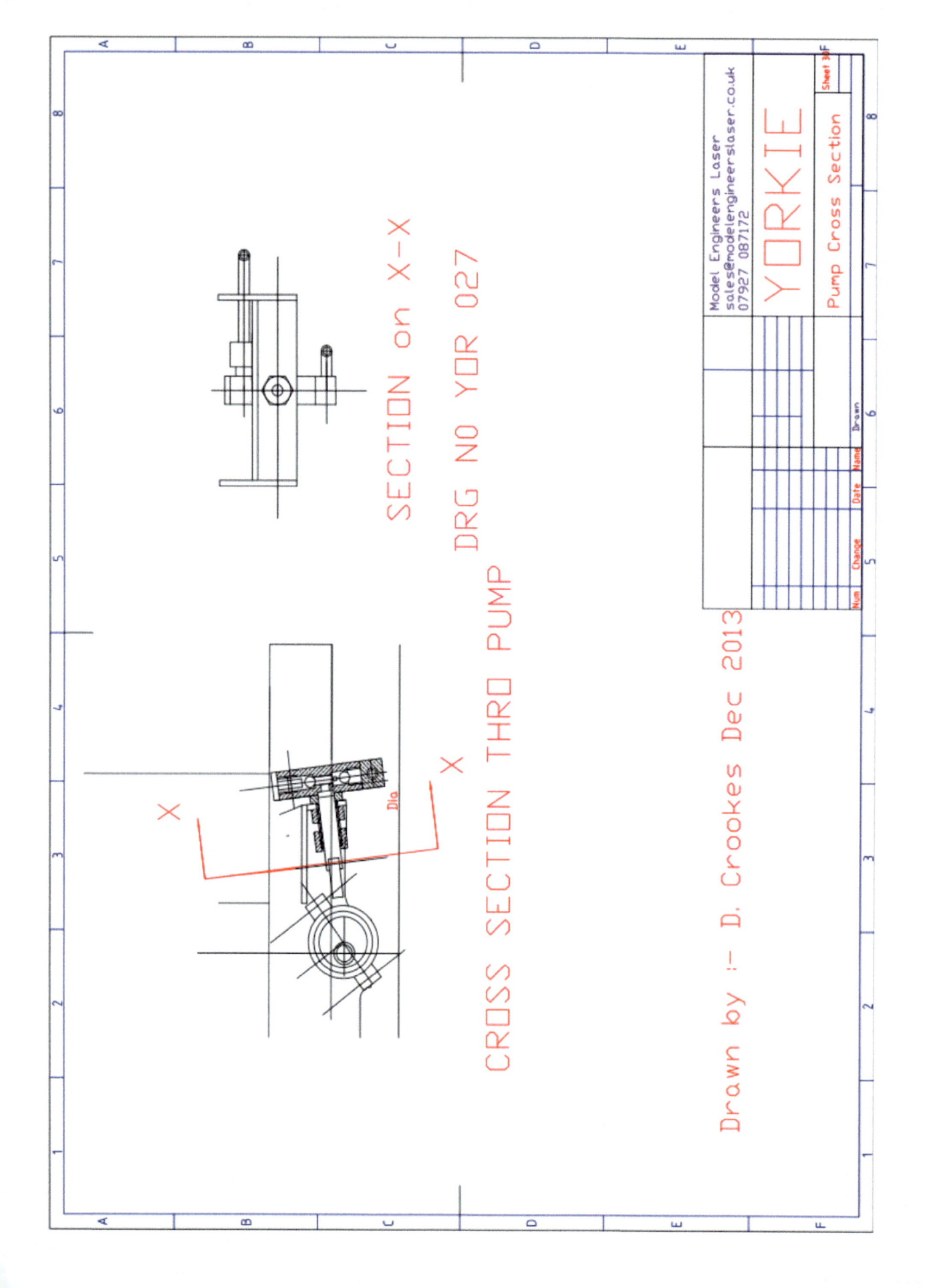

Model Engineers Laser
sales@modelengineerslaser.co.uk
07927 087172

YORKIE

Pump Cross Section

BUFFER FOOTPLATE

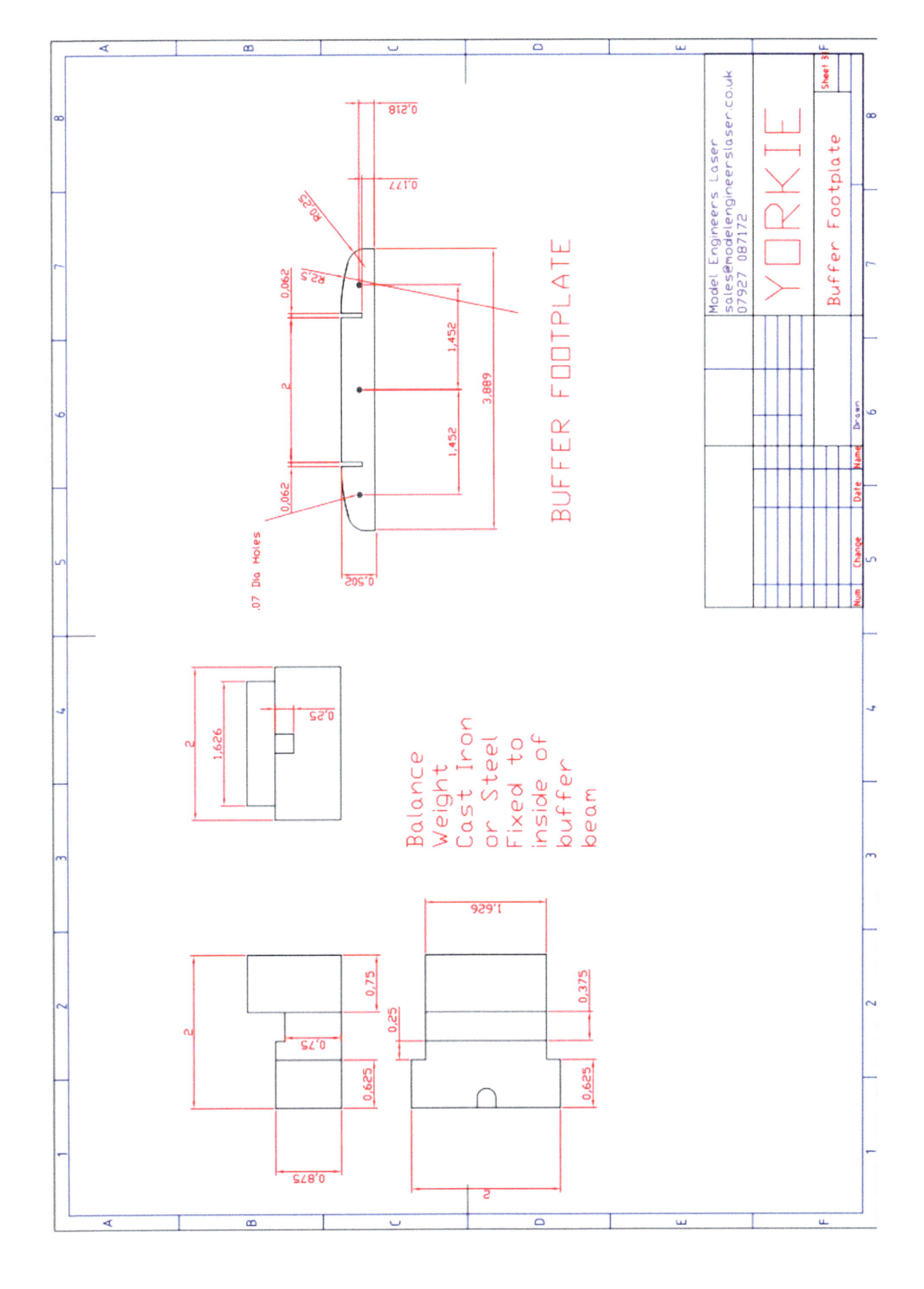

YORKIE
Buffer Footplate

Model Engineers Laser
sales@modelengineerslaser.co.uk
07927 087172

.07 Dia Holes

Balance Weight Cast Iron or Steel Fixed to inside of buffer beam

REACH ROD

PIVOT PIN

DRG No YOR 030

Drawn by :- D. Crookes Dec 2013

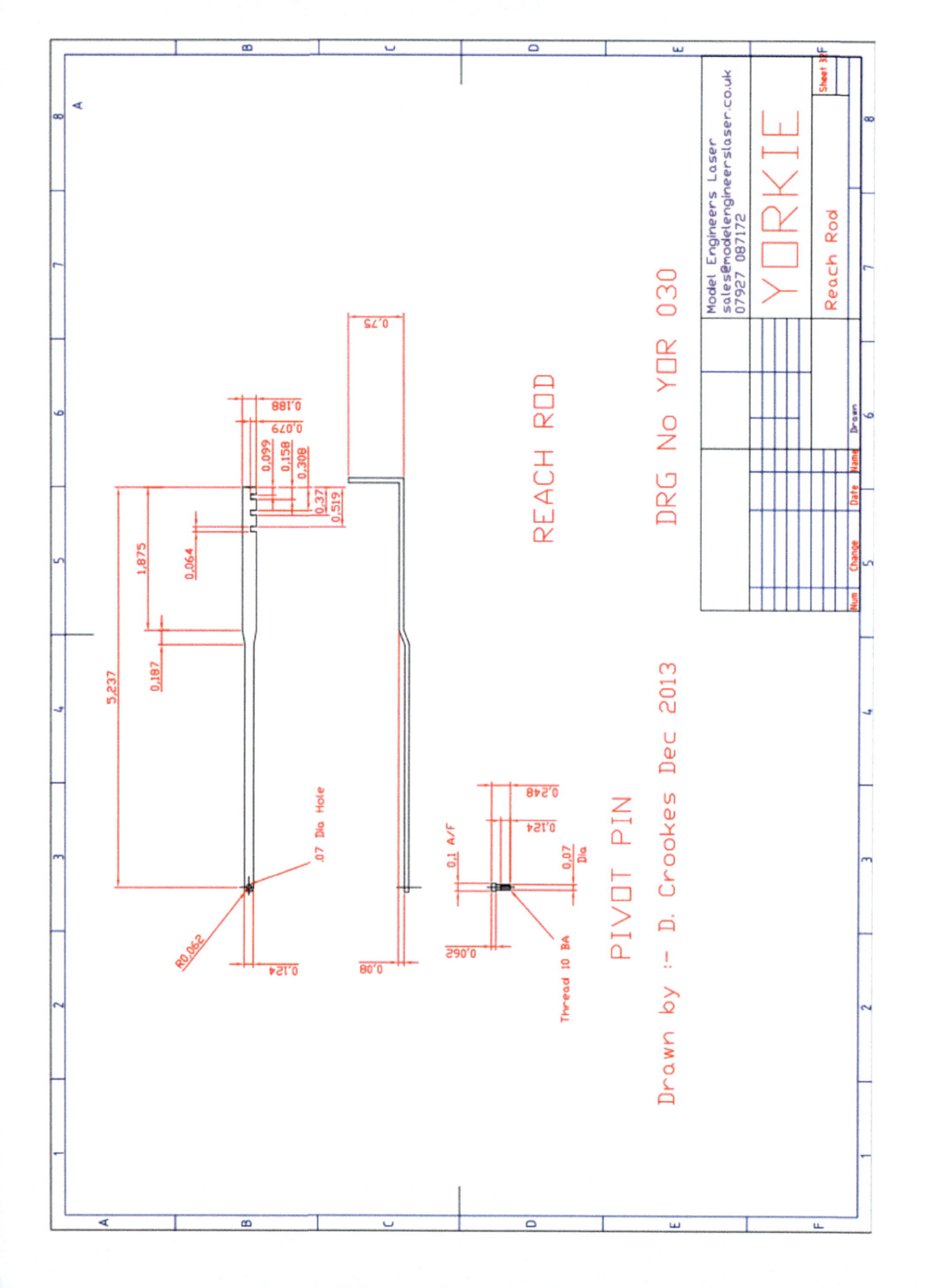

YORKIE

Model Engineers Laser
sales@modelengineerslaser.co.uk
07927 087172

Reach Rod

CAB ROOF SUPPORTS

Material :- Brass & St St

4 off Req'd   Drg No YOR 035

- ⅙₆ Dia St St Loctite in position
- 2 Holes Tap 10 BA
- Drill & Tap 10 BA x ¼ Deep
- 0.188 Sq Brass
- 0.375
- 2.157
- 3.562
- 1.78
- 0.34

STEPS

1 off As Drawn
1 off Opp Hand

Material :- Brass

DRG No YOR 032

Drawn by :- D. Crookes Dec 2013

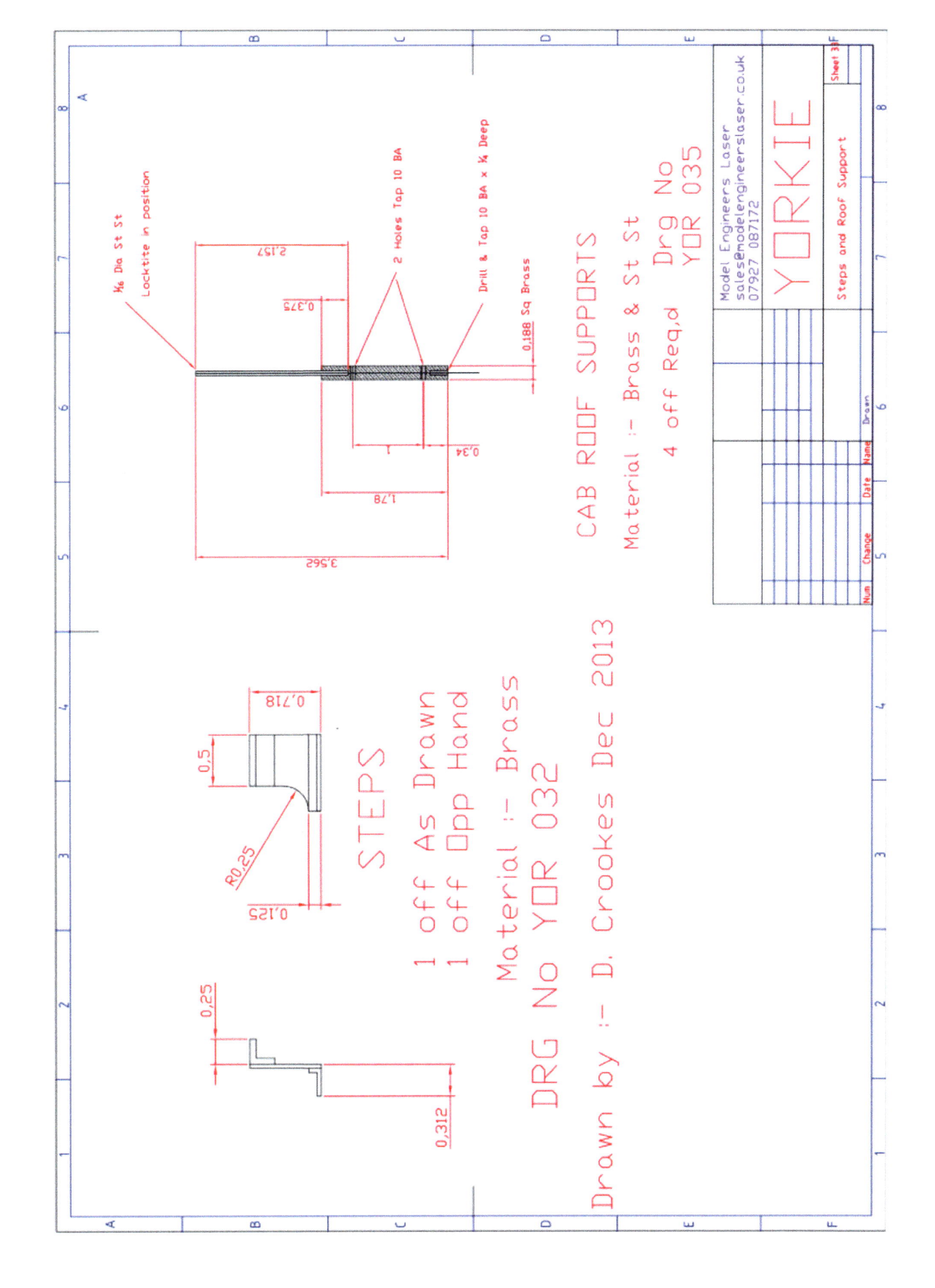

- 0.718
- 0.5
- R0.25
- 0.125
- 0.25
- 0.312

Model Engineers Laser
sales@modelengineerslaser.co.uk
07927 087172

YORKIE

Steps and Roof Support

Sheet 3 F

Drill & Tap 4.5mm x .5mm
Recess .218 Dia x .03 Deep

Drill & Tap 3/16 x 40

0.866
0.5
0.062
0.187
3
0.124
0.062
0.111
0.25

Thread 6 BA

Note ! All Silver Solder Construction
To Be Hydraulically Tested To A Min 300 P.S.I.

GAS TANK
Material :- Copper Tube & Brass

Drawn by :- D. Crookes Dec 2013

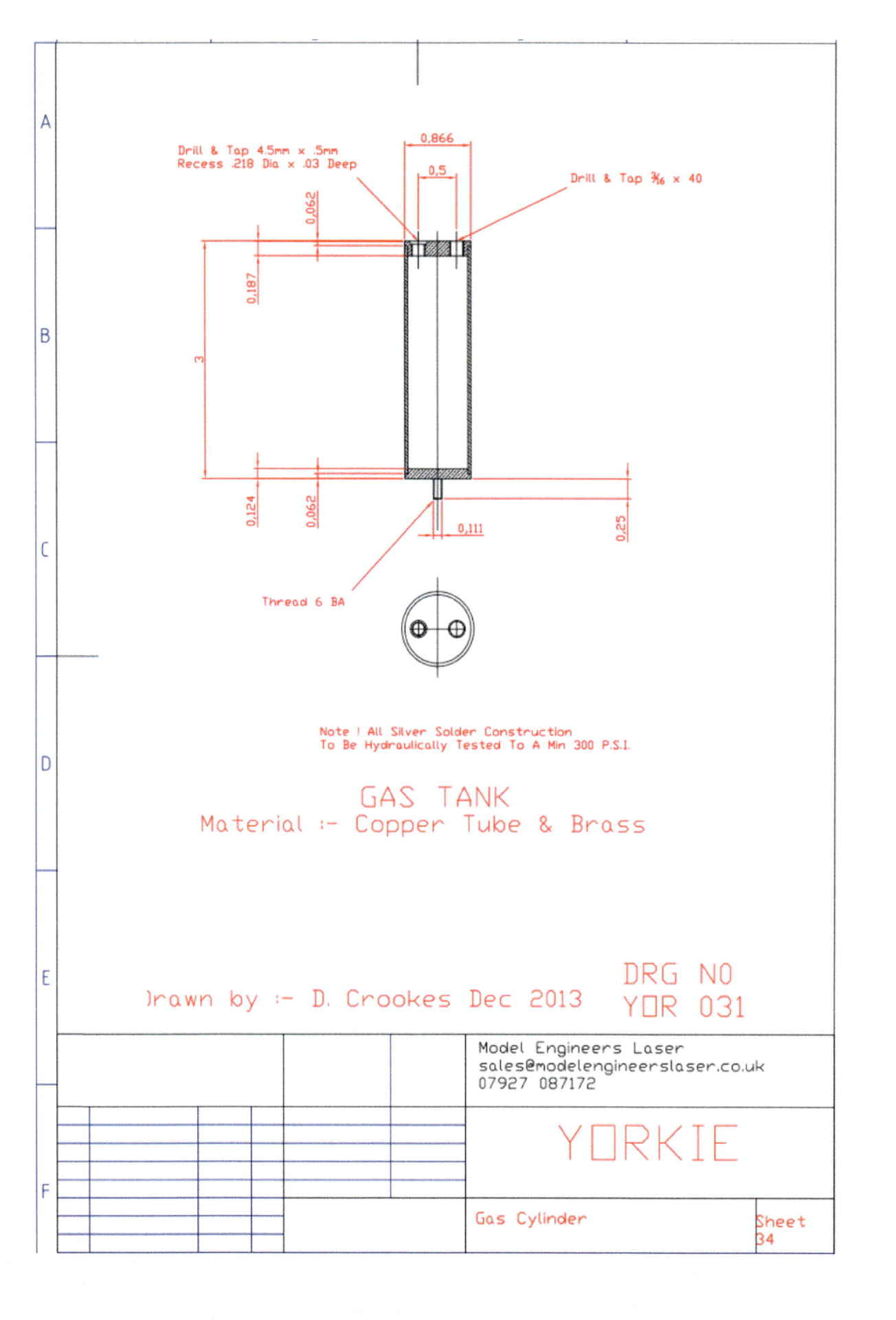

DRG NO
YOR 031

Model Engineers Laser
sales@modelengineerslaser.co.uk
07927 087172

YORKIE

Gas Cylinder

Sheet 34

LOCKNUT

GAS VALVE
Material :- Brass

3mm St ST Screwed Rod
Cone end @ 30 deg

1/16 Dia Holes

Drawn by :- D. Crookes Dec 2013

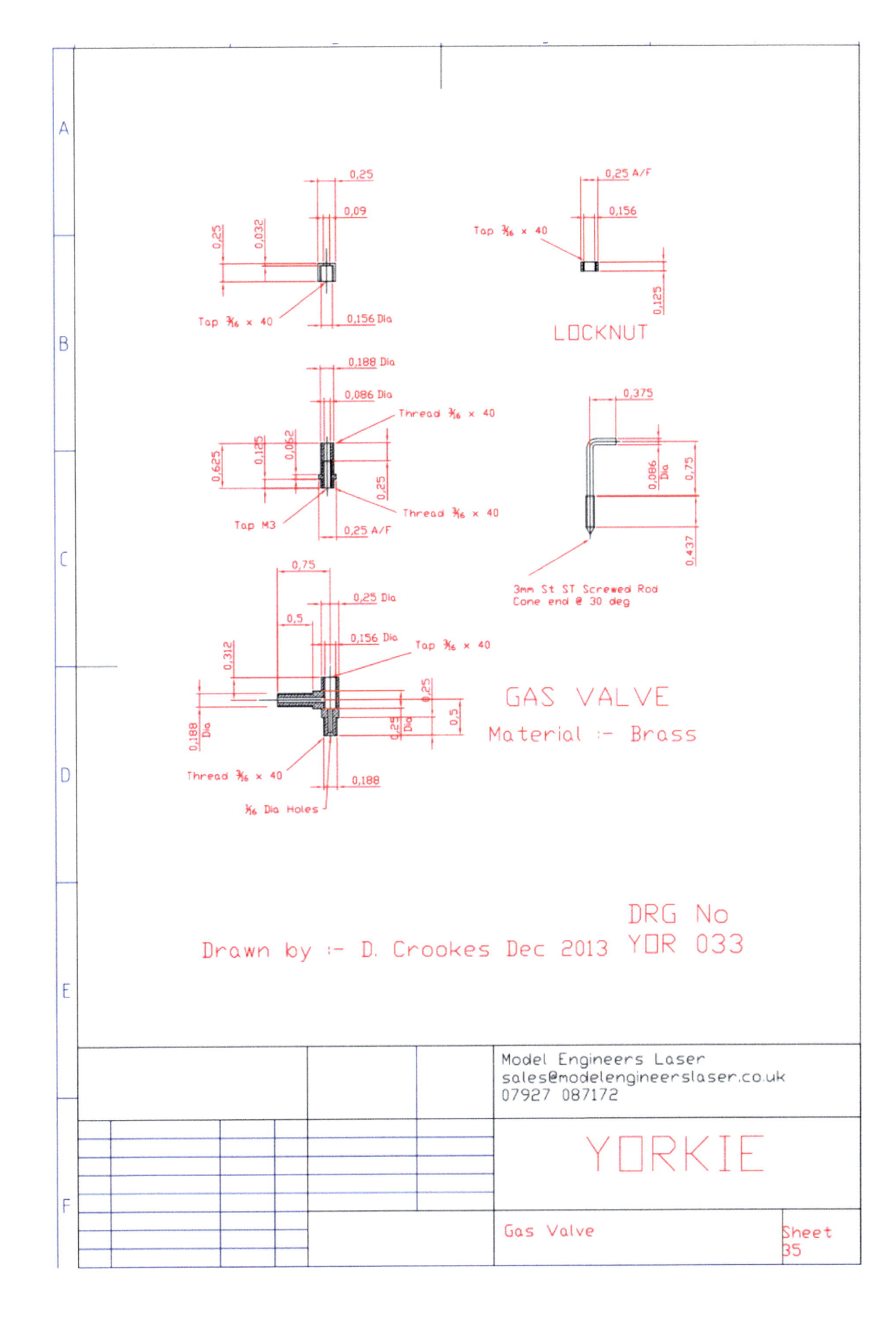

DRG No YOR 033

Model Engineers Laser
sales@modelengineerslaser.co.uk
07927 087172

YORKIE

Gas Valve

Sheet 35

CAB BACK
Material :- 1mm Brass

DRG No YOR 034

Drawn by :- D. Crookes Dec 2013

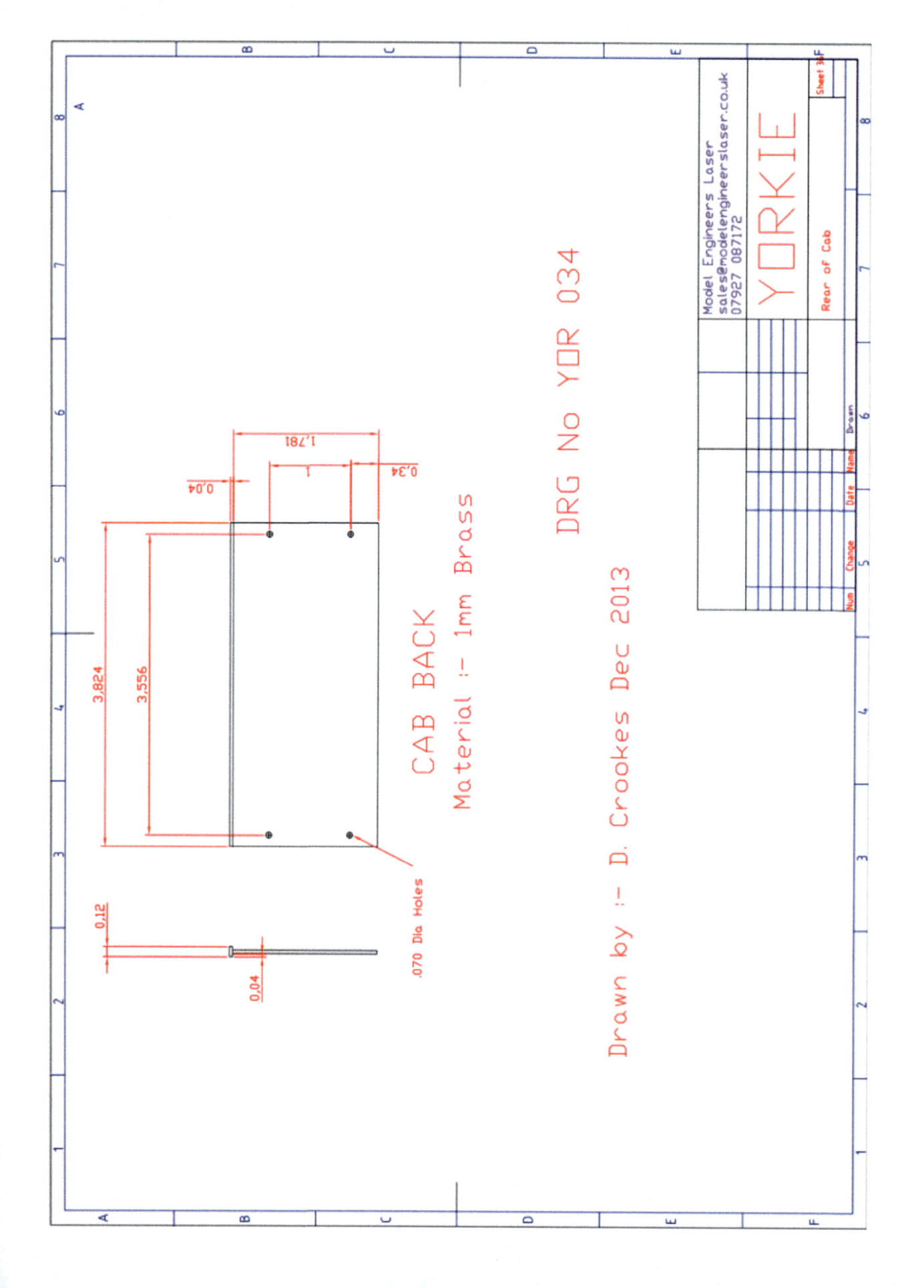

.070 Dia Holes

3.824
3.556
1.781
0.04
0.34
0.12
0.04

YORKIE
Rear of Cab

Model Engineers Laser
sales@modelengineerslaser.co.uk
07927 087172

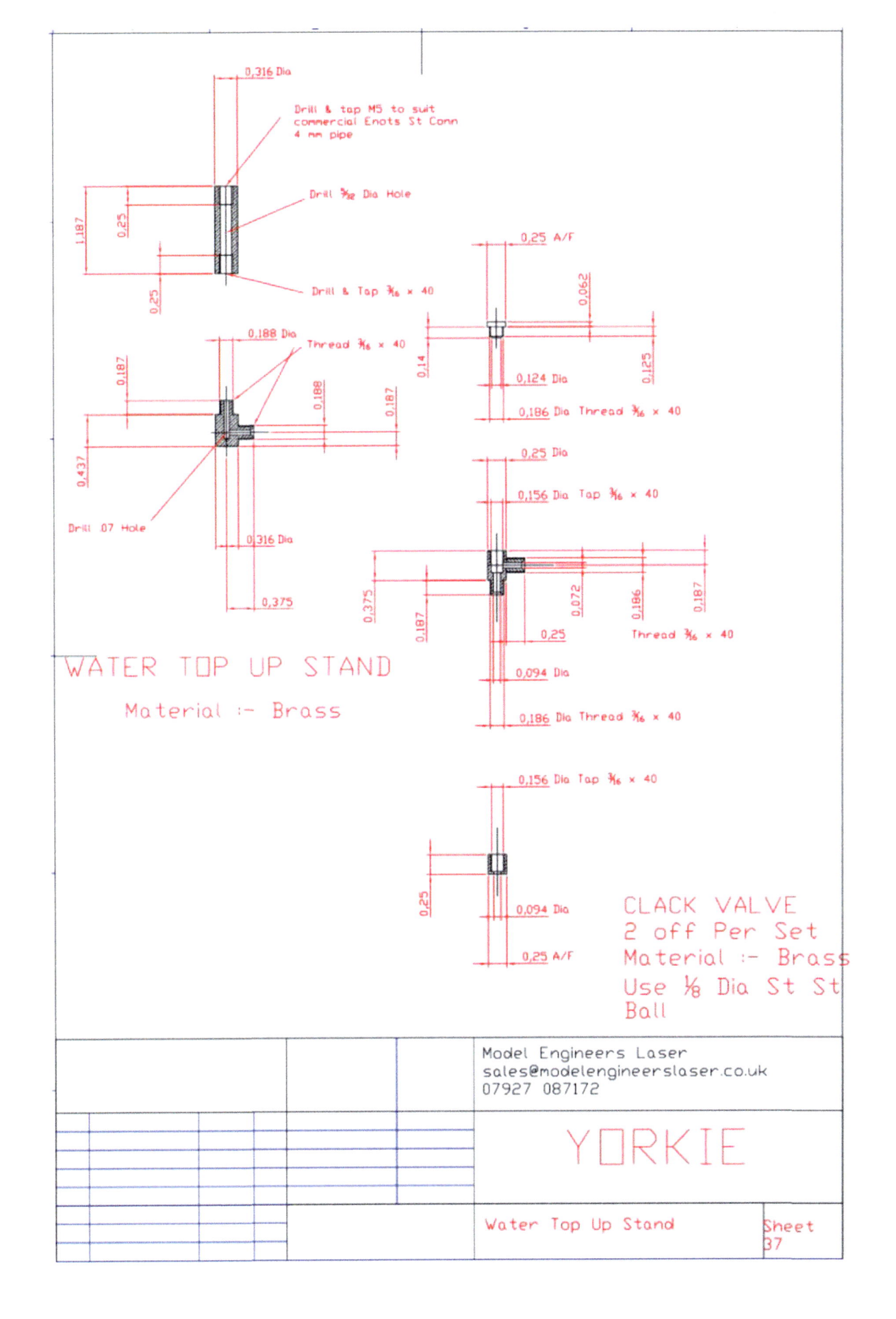

WATER TOP UP STAND
Material :- Brass

CLACK VALVE
2 off Per Set
Material :- Brass
Use ⅛ Dia St St Ball

Model Engineers Laser
sales@modelengineerslaser.co.uk
07927 087172

YORKIE

Water Top Up Stand

Sheet 37

DOME
Material :- M.S.

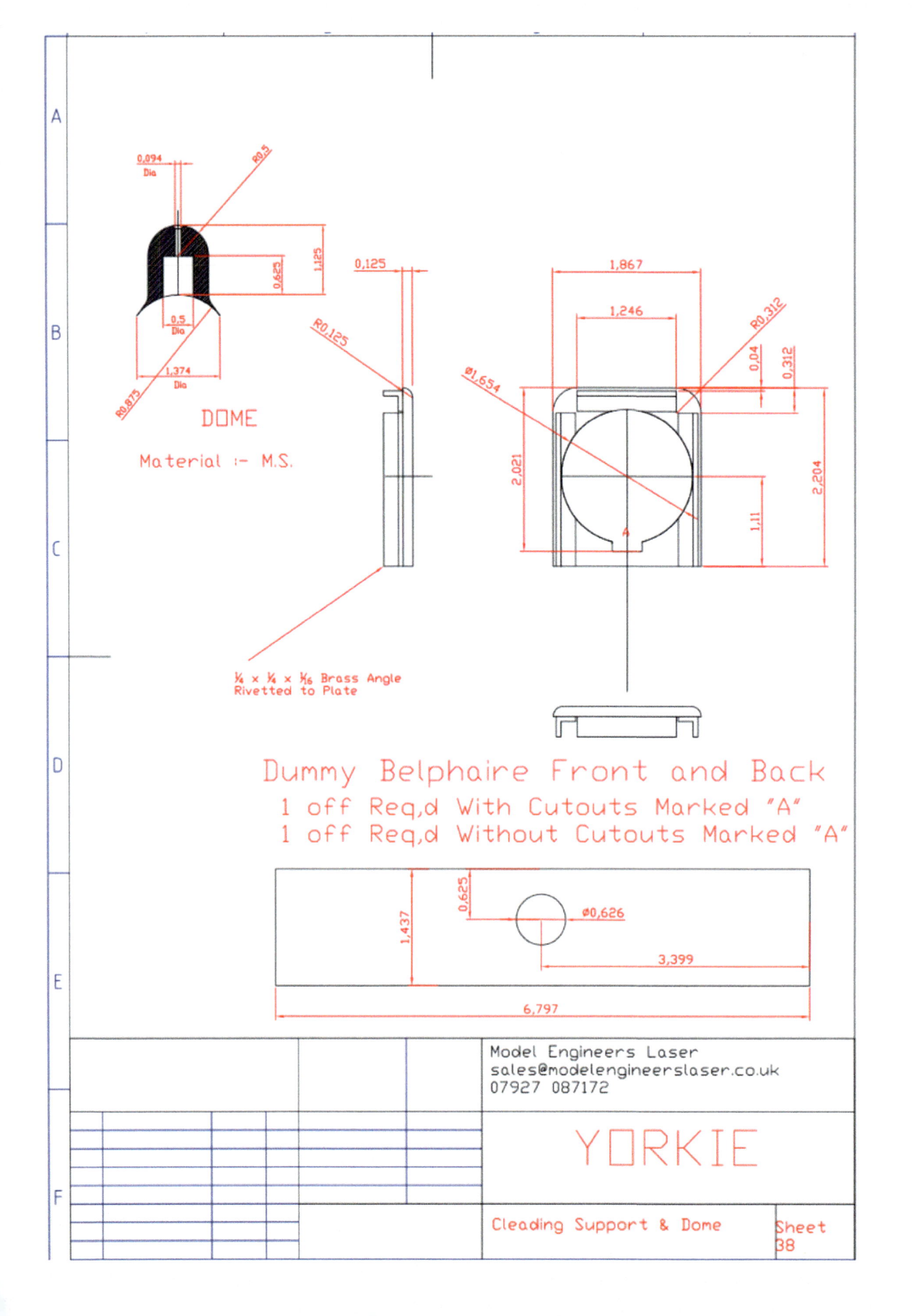

¼ × ¼ × 1⁄16 Brass Angle Rivetted to Plate

Dummy Belphaire Front and Back
1 off Req,d With Cutouts Marked "A"
1 off Req,d Without Cutouts Marked "A"

Model Engineers Laser
sales@modelengineerslaser.co.uk
07927 087172

YORKIE

Cleading Support & Dome

Sheet 38

SMOKEBOX DOOR
Material :- M.S.

.090 Dia Hole
0,135
0,03
Ø1,652
C/L of Hinge
R3,083
R0,03
1,2
1,3
0,3

Door Hinge
0.5mm Brass 2 off

Ø0,039
1,187
0,047
0,135
0,25
0,094
Three Off
0,09
0,06
0,184

Closer 1.5mm Brass Rivet to back of smoke box front
Tap 8 BA
1,5
0,25

DART
Material :- St St

0,186
0,375
0,107
Thread 10 BA
Thread 8 BA
0,07
0,502
0,09
0,156
10°

Commercial Handwheel 7/16 Dia Approx

Drawn by :- D. Crookes Dec 2013
DRG No YOR 038

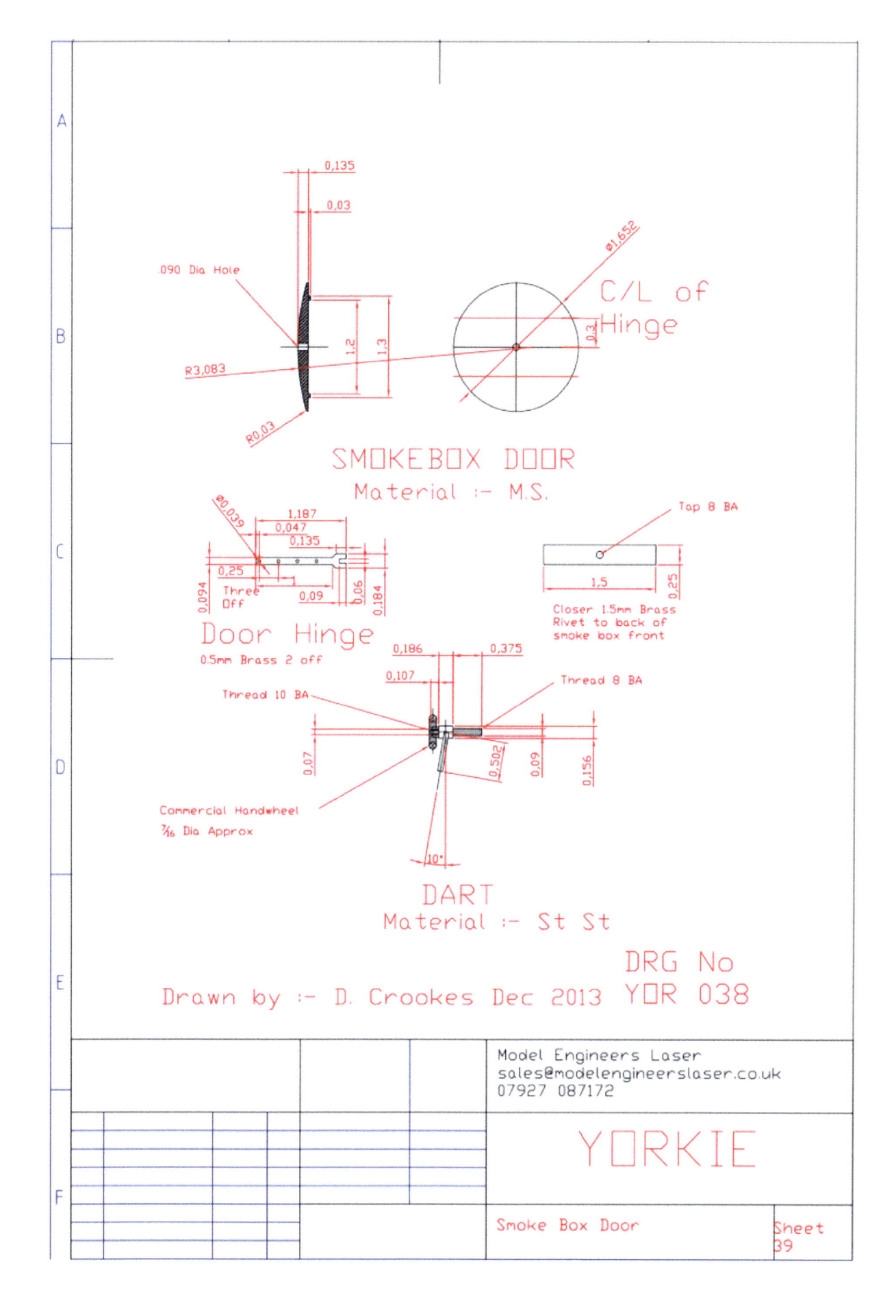

Model Engineers Laser
sales@modelengineerslaser.co.uk
07927 087172

YORKIE

Smoke Box Door

Sheet 39

CAB ROOF SUPPORT SOCKET

Material :- Brass
4 off Req.d

0.094 Dia
0.125
1/16 Dia Hole x 3/16 Deep
0.25
0.15 Dia

CAB ROOF
Material :- 1mm Brass

3.904
R7.156
0.174
0.174
0.375
0.625
0.5
3/64 Dia Holes
3.625
2.375
1/4 Dia Hole
1.375
0.687
1/16 Dia Holes
0.25

Drawn by :- D. Crookes Dec 2013
DRG No
YOR 039

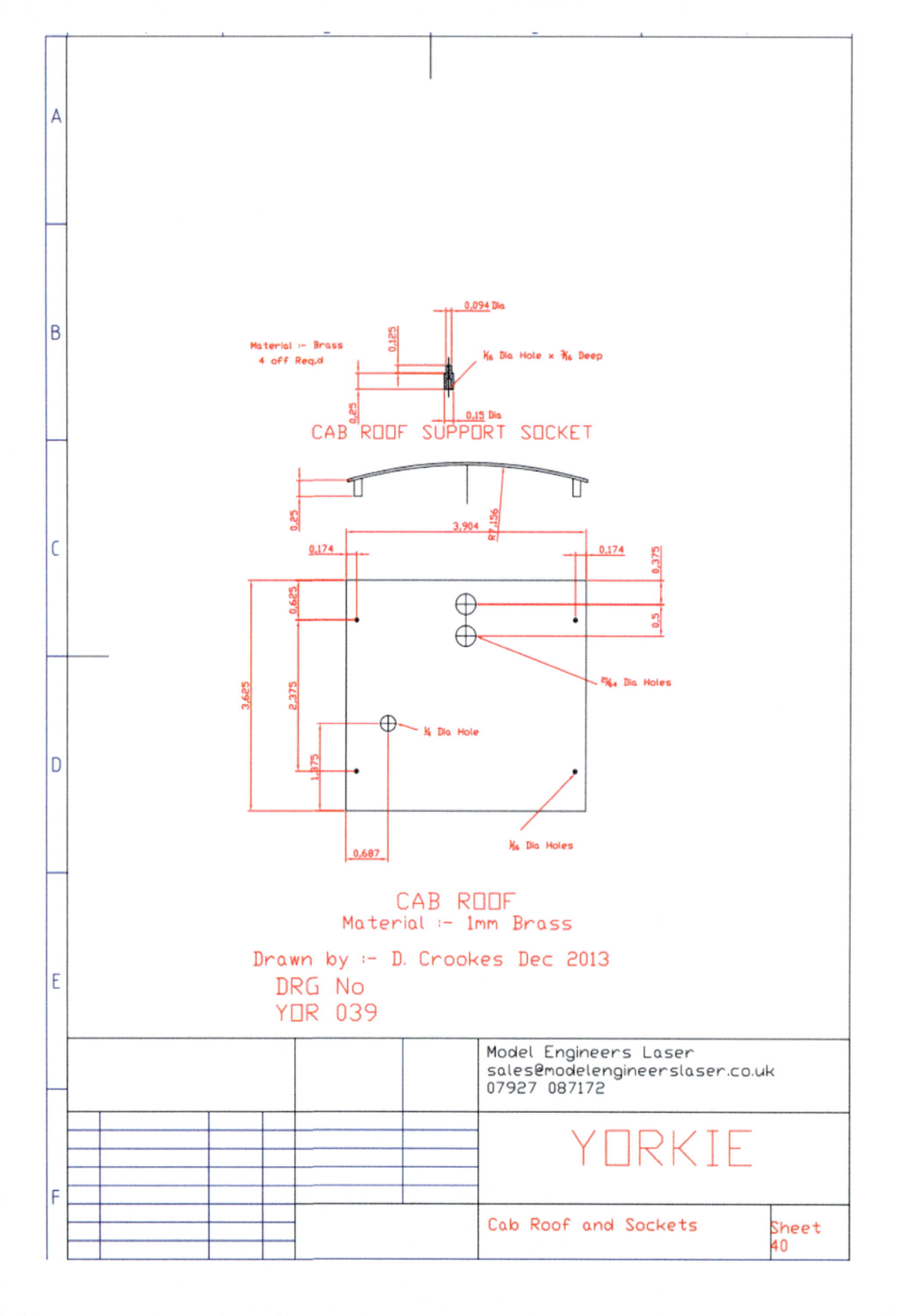

Model Engineers Laser
sales@modelengineerslaser.co.uk
07927 087172

YORKIE

Cab Roof and Sockets

Sheet 40

SECTION THRO STEAM MANIFOLD SECTION THRO EXHAUST MANIFOLD

Drawn by :- D. Crookes Dec 2013   DRG No YOR 041

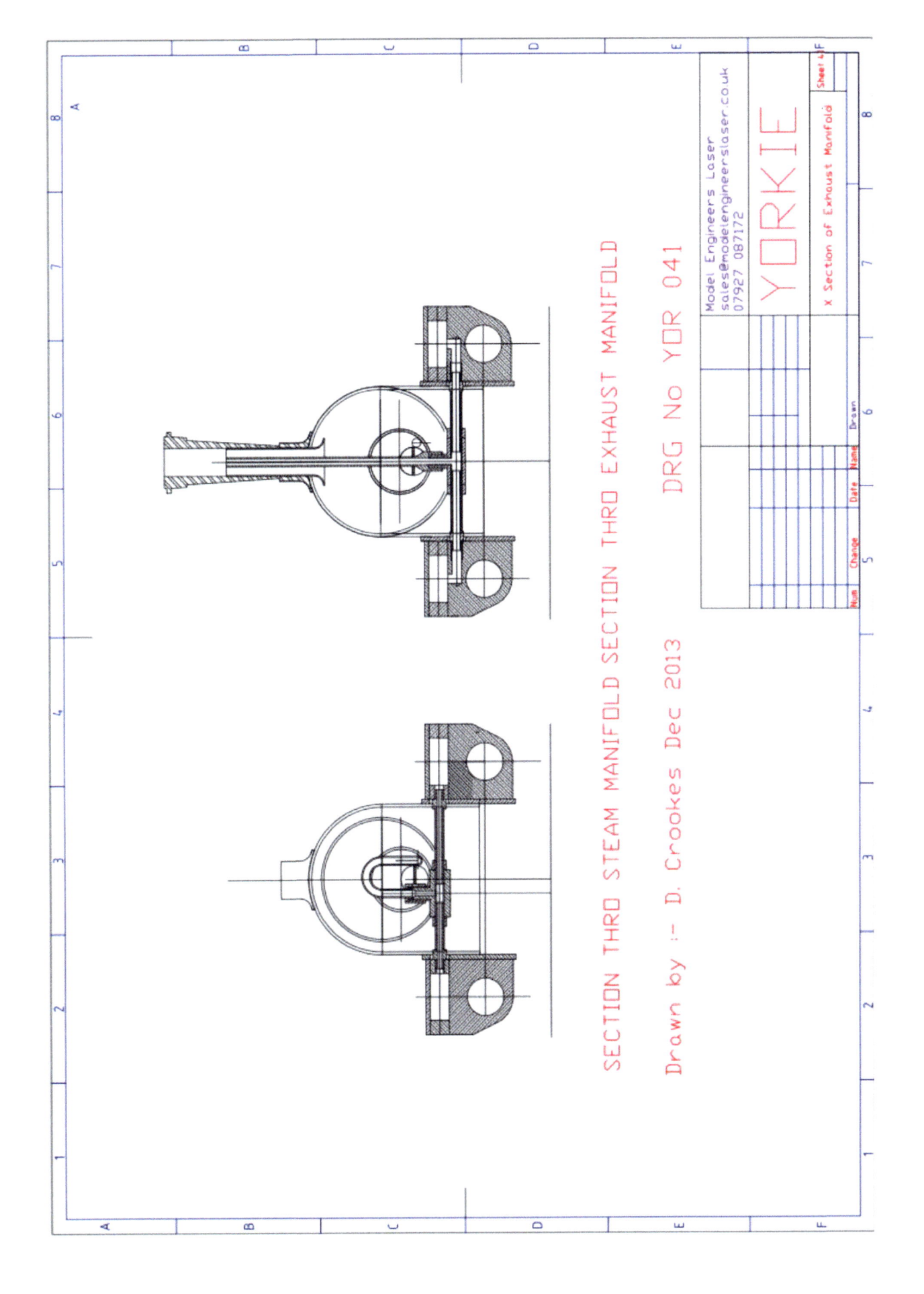

YORKIE

Model Engineers Laser
sales@modelengineerslaser.co.uk
07927 087172

x Section of Exhaust Manifold

Sheet 4F

"X"
0.125
0.125
Thread both ends ⅛ × 40
0.188 A/F
0.125 Dia
0.064 Dia

1 Off Where "X" = 0.9
1 Off Where "X" = 1.25

Thread 3/16 × 40
0.188 Dia
.070 Dia hole
0.25 A/F
0.365
Drill & tap ⅛ × 40
0.625

0.125
0.25 A/F
Drill & tap ⅛ × 40

2 Off Req'd

STEAM MANIFOLD
Material :- Brass

Drawn by :- D. Crookes Dec 2013

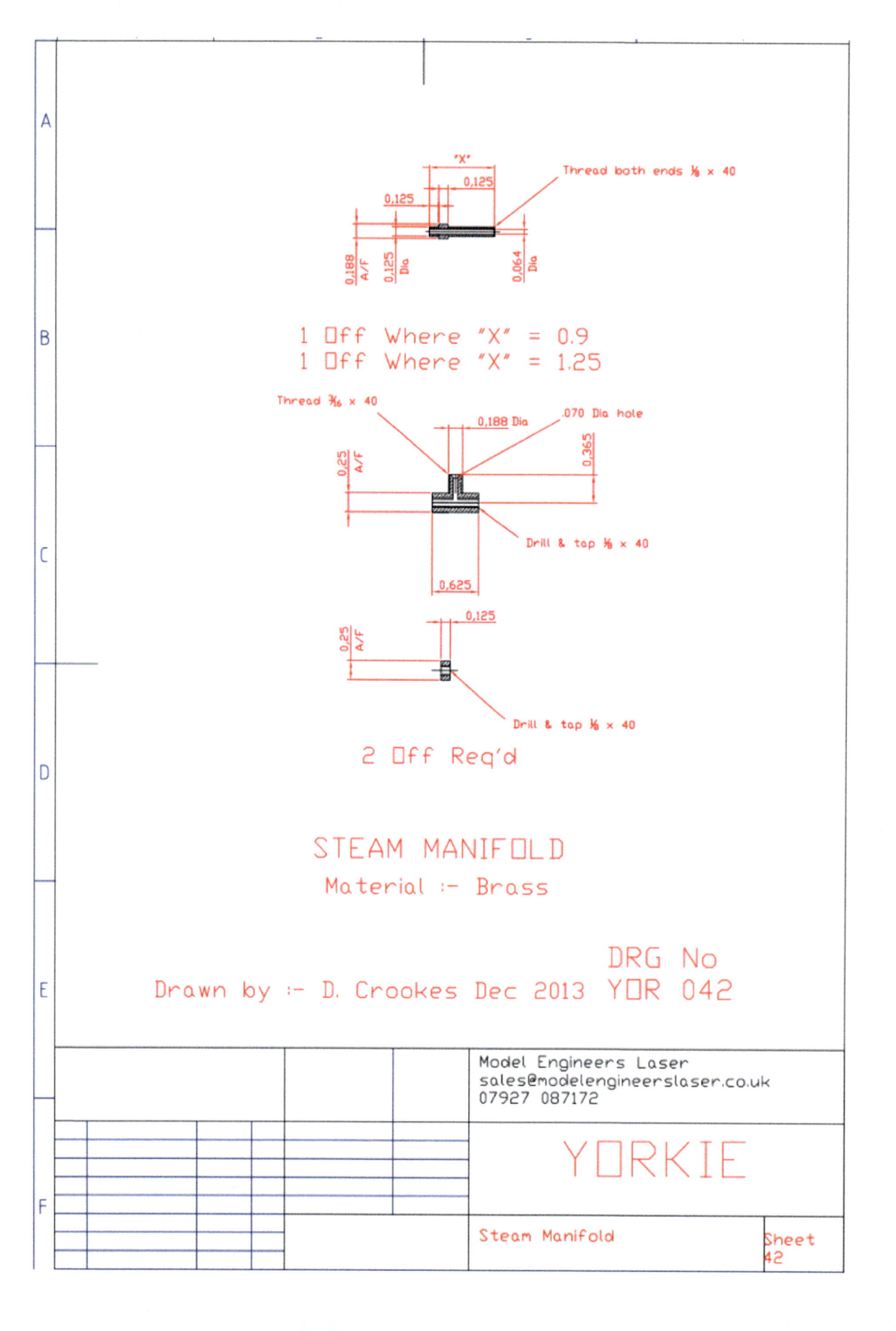

DRG No YOR 042

Model Engineers Laser
sales@modelengineerslaser.co.uk
07927 087172

YORKIE

Steam Manifold

Sheet 42

EXHAUST MANIFOLD

Material :- Brass

Drawn by :- D. Crookes Dec 2013

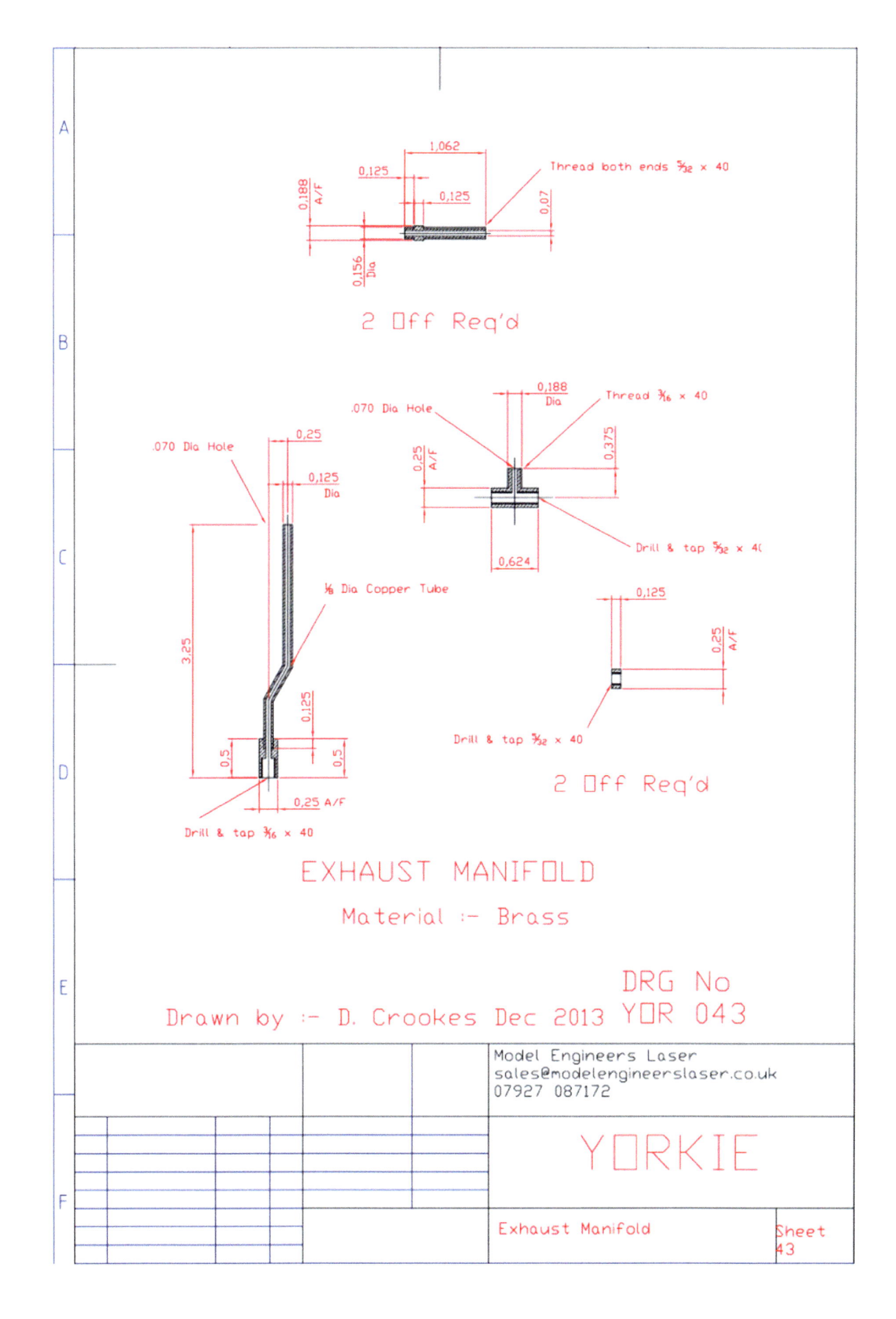

DRG No YOR 043

Model Engineers Laser
sales@modelengineerslaser.co.uk
07927 087172

YORKIE

Exhaust Manifold

Sheet 43

BOILER SUPER HEATER STEAM PIPE

DRG No YOR 051

Drawn by :- D. Crookes Dec 2013

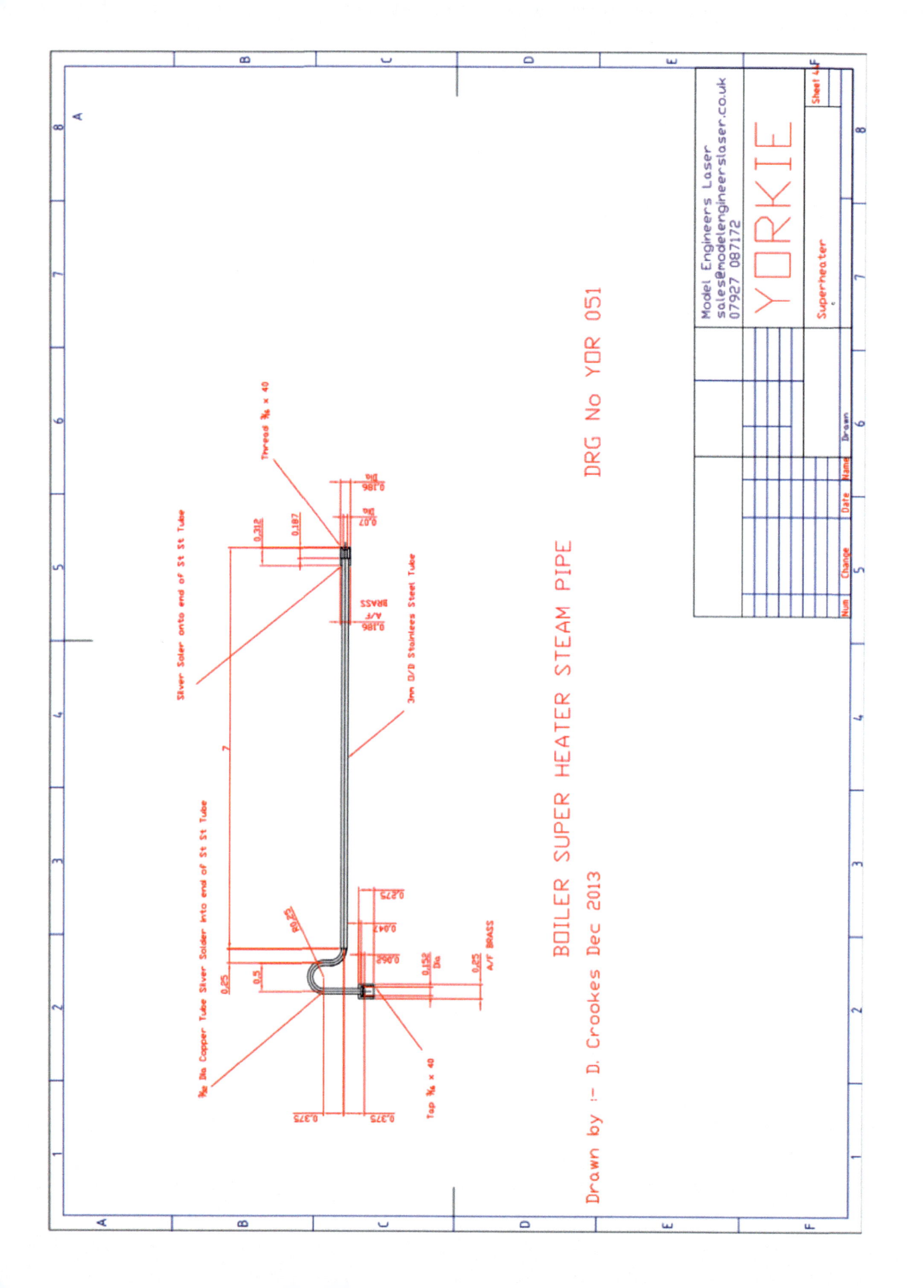

Appendix One

"Yorkie" is a popular engine that we have had many requests from people who are without access to engineering machinery, on whether we can supply wheels and cylinders, to allow them to build "Yorkie", too.

To overcome this, we have investigated the possibilities of using proprietary components. Some design work and collaboration has been done with Roundhouse Engineering and we have found that with very little modification to our design, some of their standard parts can be used in the build. There are some slight variations to our kit that will require special parts from us or make your own as per the drawings that follow, these are parts such as modified frames and rods: if you are going to build this version, you will need to specify this to us when ordering your "Yorkie".

The parts that you will need to buy from Roundhouse will require assembly as per Roundhouse instructions with the following exceptions.

1. Discard the eccentric rods provided in the Roundhouse kit as both of them are too long and you need to use the special ones that we provide. These are not joggled, however, the return crank end of the rod is of deeper section than the expansion link end.
2. The middle driving wheels are too close to the Expansion Link Bush to fit the Starlock Washer, as is shown in the instructions. You need to make the tube shown in the drawings to space it clear.
3. Roundhouse models are made for Left Hand drive, but "Yorkie" is Right Hand drive, so a modification is needed to the Lifting Arms.
   a) The original right hand arm needs to be fitted to the left side of the engine.

b) The original left side needs it's Reversing Arm removing, bending the opposite way, and then soldering back on, to flip it's orientation.

The parts you will be required to purchase from Roundhouse are as follows:

- Disc wheels, Part Number WD – you will need 4 flanged and 2 unflanged
- Cylinder set, Part number C
- Walschaerts valve gear, Part Number WVG
- Axle, Part number AXLE – you will need 3
- (Optional) Crosshead and Combination Lever, Part Number CLCH
- Lubricator, Part Number LUB

On the following pages you will find the modified drawings.

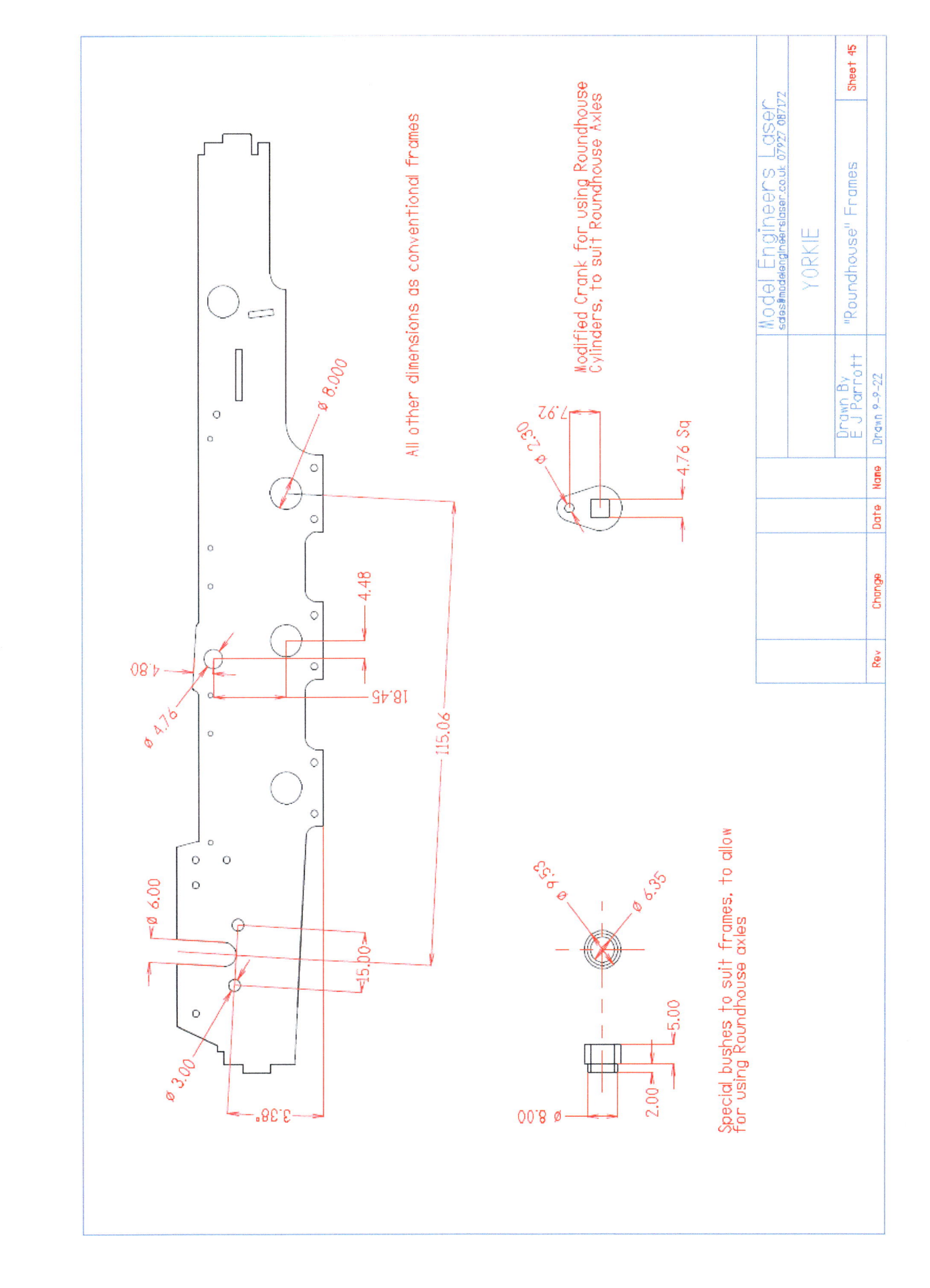

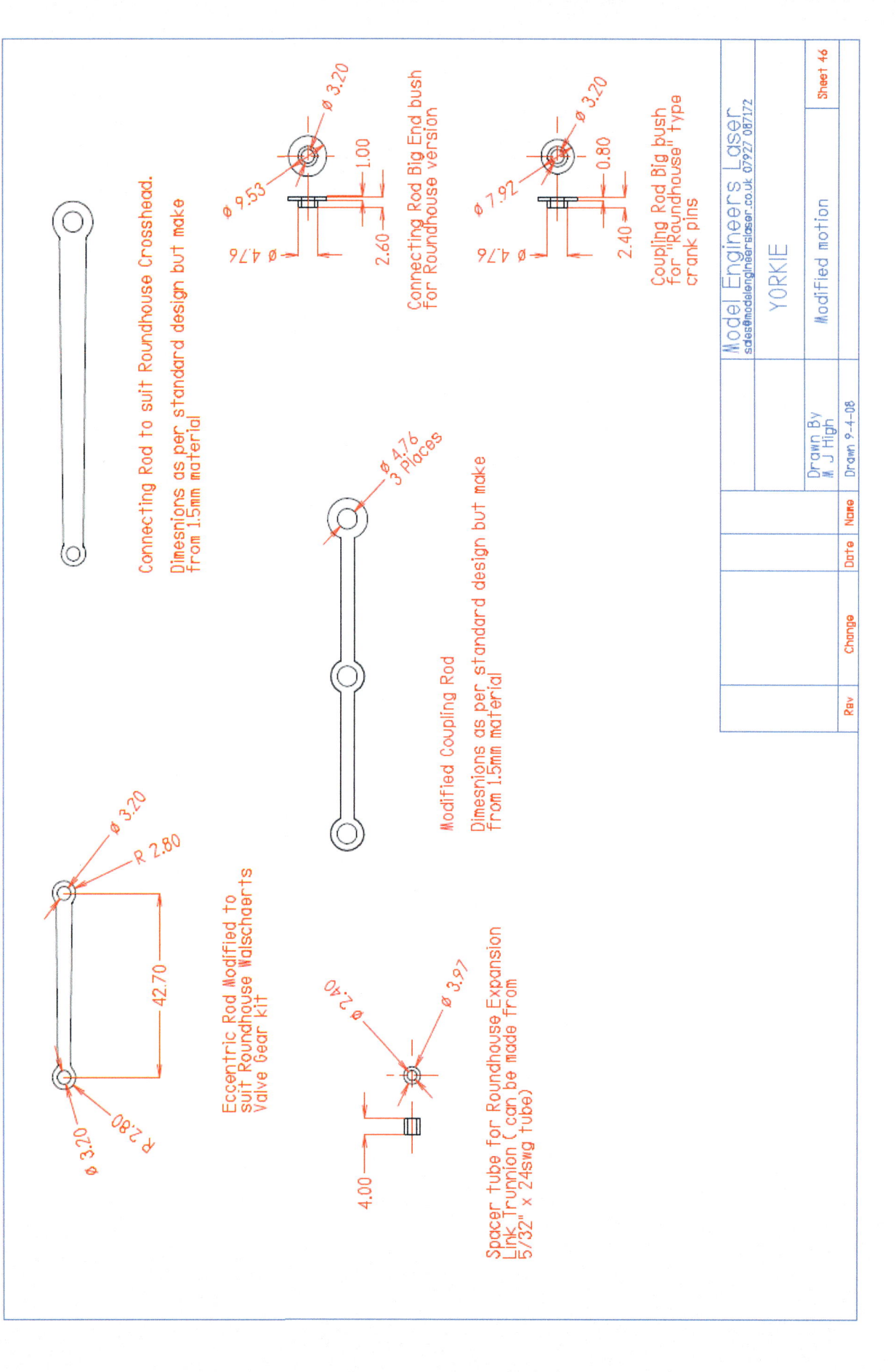

Published by Northstar Support

Printed in Great Britain
by Amazon